THE ISLANDER

The extraordinary adventures of the man who wanted to be Robinson Crusoe

Being a true account of his adventures on

Cocos Island
5 30′N, 87 W
300 miles north of the Galapagos Islands, Mid-Pacific

Robinson Crusoe Island
34 S, 79 W
Juan Fernandez Archipelago, South Pacific

Tuin Island
10 10′S, 142 E
Torres Strait, between Australia & Papua New Guinea

Badu Island
Three kilometres north of Tuin

The Islander

Gerald Kingsland

NEW ENGLISH LIBRARY
Hodder and Stoughton

To all those who helped me
make this book possible

First published in Great Britain in 1984 by
New English Library

First NEL Paperback Edition November 1985
Fourth impression 1986

British Library C.I.P.

Kingsland, Gerald
  The islander.
  1. Islands of the Pacific – Description and
  travel
  I. Title
  919'.04   DU23.5

  ISBN 0 450 05853 0

Printed and bound in Great Britain for
Hodder and Stoughton Paperbacks, a
division of Hodder and Stoughton Ltd.,
Mill Road, Dunton Green, Sevenoaks,
Kent (Editorial Office: 47 Bedford
Square, London, WC1B 3DP) by
Cox & Wyman Ltd., Reading.

Contents

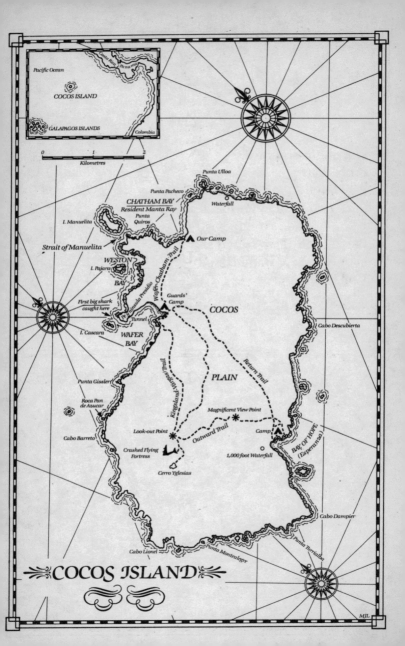

COCOS ISLAND

Pacific Ocean

COCOS ISLAND

GALAPAGOS ISLANDS

Colombia

0   1   2
Kilometres

Punta Ulloa

Punta Pacheco

CHATHAM BAY

Waterfall

Resident Manta Ray

Punta Quiros

I. Manuelita

Our Camp

Strait of Manuelita

WESTON

I. Pajara

BAY

Guards' Camp

COCOS

First big shark caught here

Cabo Descubierta

Tunnel

I. Cascara

WAFER

BAY

PLAIN

Punta Gissler

Roca Pan de Azucar

Magnificent View Point

Look-out Point

Camp

Cabo Barreto

Outward Trail

BAY OF HOPE

(Esperanza)

Crashed Flying Fortress

1,000 foot Waterfall

Cerro Yglesias

Cabo Dampier

Punta Turrialba

Cabo Lionel

Punta Montealegr

MJL

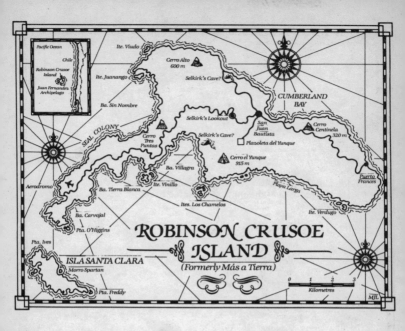

ROBINSON CRUSOE ISLAND
(Formerly Más a Tierra)

Pacific Ocean
Chile
Robinson Crusoe Island
Juan Fernandez Archipelago

Ite. Viudo
Cerro Alto 600 m
Ite. Juanango
Selkirk's Cave?
Ba. Sin Nombre
CUMBERLAND BAY
Selkirk's Lookout
SEAL COLONY
San Juan Bautista
Cerro Tres Puntas
Cerro Centinela 320 m
Selkirk's Cave?
Plazoleta del Yunque
Ba. Villagra
Cerro el Yunque 915 m
Aerodromo
Ba. Tierra Blanca
Ite. Vinillo
Puerto Frances
Playa Larga
Ba. Carvajal
Ites. Los Chamelos
Pta. O'Higgins
Ite. Verdugo
Pta. Ives
ISLA SANTA CLARA
Morro Spartan
Pta. Freddy

0 1 2 3
Kilometres

MJL

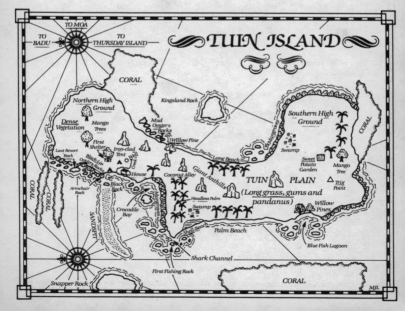

TUIN ISLAND

TO BADU
TO MOA
TO THURSDAY ISLAND

CORAL

Northern High Ground
Kingsland Rock
Southern High Ground
Dense Vegetation
Mango Trees
Mud Oyster Rocks
First Shelter
Willow Pine
Iron-clad Tent
Swamp
Last Resort Rock
Black-lip Oyster Rocks
Creek
Sweet Potato Garden
House
Coconut Alley
Giant Anthills
Mango Tree
Armchair Rock
Black Rock
Mangroves
TUIN PLAIN
Brig Point
Crocodile Bay
Headless Palm
(Long grass, gums and pandanus)
Willow Pines
Swamp
SANDSPIT
Shell Beach
Palm Beach
Blue Fish Lagoon
CORAL
CORAL
Shark Channel
First Fishing Rock
Snapper Rock
CORAL

MJL

Illustrations

The western arm of Wafer Bay from a hill above the guards' camp
   Photograph by the author
Roddick in Chatham Bay   *Photograph by John Froy*
Pajara Island, with Manuelita Island in the background   *Photograph by the author*

Wafer's lone high waterfall   *Photograph by the author*

The creek in Wafer Bay where Benito Bonito supposedly hid his dubloons   *Photograph by the author*
Anne Hughes, making a fishing net from her mosquito net to catch sardines   *Photograph by the author*

Roddick with the remains of the crashed Flying Fortress   *Photograph by Richard Evans*

Lucy and Gerald Kingsland on board the *Torres Strait Islander*, going to Tuin   *Photograph by Jackie Mott*
The author spearing crab for bait in Crocodile Bay   *Photograph by Lucy Kingsland*

The author with his newly cleaned shotgun   *Photograph by Lucy Kingsland*

Surveying saplings to cut for the first shelter   *Photograph by Lucy Kingsland*
The author in his office, as the first rough shelter became known   *Photograph by Lucy Kingsland*

Ronald Lui, his children and one of their friends   *Photograph by Lucy Kingsland*
Badu women preparing a green turtle   *Photograph by Lucy Kingsland*

The author is grateful to the *Daily Telegraph* Colour Library for permission to reproduce the pictures taken on Tuin Island. He also wishes to thank John Froy and Richard Evans for permission to reproduce their pictures taken on Cocos Island.

Prologue

SHE SWUNG her long, tanned legs into the signal-red E-type. I closed the door with a soft clunk, got in behind the steering wheel. As always, Carol looked a million dollars: white-blonde straight hair, curling softly inwards at the ends to brush her blue silk dress below the shoulders; oval face impeccably made up. People stared at her in the restaurant; men with longing, women with jealousy.

'So this is the last time I'll see you for quite a while,' she said. 'You really are going.'

'I told you I was,' I said. 'It's all arranged.' I patted the padded top of the fascia. 'I'll be sorry to see this bitch go. She has done her level best to kill me.'

I turned the ignition key. The twelve cylinders fired with a muffled growl. 'I love that sound,' she said. I let in the clutch. It was a clear, starlit June night. I parked on high ground, overlooking the lights of the North Wales coastline.

'It's incredible, really,' she said. 'I have loved you for seven years ever since our eyes first met in that hotel bedroom. I was the scatter-brained model and you were the wealthy but very much married publisher. God, I really fancied you.'

'It was just my money,' I said, but we both knew that wasn't true.

'Why *did* you have to give it all up? You had it made, and you had me. But you had to be a wine grower in Italy. Now you've given that up for this stupid, dangerous idea. Why?'

'I am searching for something,' I said. 'I think I may have found it.'

Five weeks later I was six thousand miles away in the Pacific Ocean.

Part One

Cocos Island

5 30'N, 87 W

Three hundred miles north of the
Galapagos Islands, Mid-Pacific

One

GUN-BOAT CAPTAINS of Costa Rica's maritime police had achieved a certain notoriety for getting lost. It was with more than a casual interest that I read newspaper accounts of their aberrant navigations. I was waiting in San José, the country's capital, for one of them to take me three hundred miles out into the Pacific Ocean. The last captain to plot a course to my destination – a mere speck on the atlas – had missed the tiny island completely. And he'd had Costa Rica's newly-elected president on board.

As I watched the mainland slip below the horizon astern, the thought, not unnaturally, occurred that this particular Motor Torpedo Boat, with such a lesser passenger as me, could end up anywhere! It was impossible to distinguish captain, helmsman and crew. All were bare-footed and attired in colourful shorts and singlets – and crowded on to the bridge. Everyone tried his hand at the wheel. Even the cook – recognisable by a chef's hat – popped up from the galley to have a go.

None of the instrument panels could be seen. Backsides covered them. But who wanted to see dials? Did it really matter that the wake curved now and then? The twin diesels were in screamful tune, the tropical sun was high in azure, and gun-boat FP407 was hurling herself through the swell, pretending to be a destroyer by throwing bow waves over her hot steel deck.

My insides were being violently shaken, not stirred, as I bumped up and down with those laughing policemen. Idiotically, I found myself laughing with them, though I hand't a clue what about. They couldn't speak English and my Spanish was limited to *siesta*, *fiesta*, *mañana* and *amigo*. With nightfall, the Pacific began to play a silly game of trying to split FP407 in half.

13

The need to sleep became urgent and I searched the boat for somewhere to put my head. The only available place was a bench alongside the galley table. Aromas of stale fish and greens intermingled to produce an aura conducive to vomiting, colourfully and skilfully enhanced by a dim green lamp. I stretched out on the hard, tossing seat. Discomfort was partially cancelled by concentrating on the adventure that awaited. My first thoughts, however, were of how ironic events had been during the interim period between the final arrangements with Costa Rica's chargé d'affaires in London and my arrival in San José. A new government had been formed and knew nothing of me and my project. At first, no one wanted to. The chargé d'affaires had left England one month ahead of me and was nowhere to be found. But I was determined to get to romantic, mysterious and uninhabited Cocos Island which had taken me six months of research to find. Suitable islands on which to be a modern-day Robinson Crusoe are very hard to come by. Those not bought by millionaires and tourist-resort speculators have been requisitioned for secret naval bases or satellite tracking and weather stations. Another important factor I'd had to consider was getting to my island once I had found it. Uninhabited islands are not in tourist brochures or on air and shipping routes.

It was quite a blow when I discovered that Cocos, after standing deserted and unwanted for centuries, had been declared a national park ten days before my plane touched down. Alvaro Ugalde, the new director general of national parks, was the one who gave me the disappointing news. In fluent English he told me: 'My government has reaffirmed suzerainty to exercise the two-hundred-mile territorial limit and stop the world's trawlers from taking the valuable tuna from Cocos's waters.'

The new president had just visited the island to erect a new plaque of ownership and hoist the Costa Rican colours.

Six police guards had been left behind to establish a radio station and patrol the waters. Evan Costa Rican trawlers had been banned. I decided to proceed, and Alvaro made arrangements for the police to take me. After all, I reasoned, it was nothing new that one of the island's two main northern bays was occupied. During the earlier part of the century, countless treasure-seekers had blown up almost every square yard of those two bays.

In 1926, when the late Sir Malcolm Campbell joined in the search, the pirate loot on Cocos was estimated to be in the region of ninety million US dollars. It included treasures of the Incas. In his book about his adventure, Campbell described Cocos as an 'eerie, awful place where unseen eyes watch your every move'. He said he had discovered a Peruvian legend that some of the Incas fled to Cocos to escape Pizarro, the Castillian bandit. Campbell's theory was that the island's sinister and formidable mountain was the home of the Lost Incas. They had dug up the pirates' gold – most of it rightfully belonged to them in any case – and took it to the mountain. That was why nothing had ever been found. But, like all the other searchers, Campbell never went further than the two bays. 'The interior is impenetrable,' he said, 'and I had neither the time nor inclination to go inland. It is truly an awesome place.'

I had both time and inclination. That mountain was to become my personal challenge – my Olympus. In that between-world of wakefulness and sleep, an absurd headline appeared: 'Lost Incas Found'. Such is the stuff dreams are made of.

But my dream on that hard bench was of someone holding me by the heels and cracking my body like a whip. I awoke. The bench was doing the cracking. FP407 was shooting up – and crashing down – with the regularity of a heartbeat. Thunderous shock waves ripped along her hull. 'Why the devil don't they slow engines?' I thought.

I wedged a shoulder and knee under the table. At least I was now going up and down with the bench instead of meeting it half way. Thankfully, sleep engulfed me once more. The cook was manufacturing breakfast smells when I was shaken at first light. *'Isla del Coco,'* I was told. This was the moment I had been waiting for: my first sight of the island. I went up the ladder to be met by warm drizzle. Everyone was jabbering in victorious ecstasy. They had made it!

There she was – my dream island, several miles away but directly ahead. A blue cottage loaf in a grey-green mist. The smaller blob on top was, of course, the mountain. Such emotions within me. I wanted to tell someone about them; but there was no one who could understand. I think, though, that those around me recognised the feeling of elation that stopped my guts from churning. FP407 had stopped bucking and was rolling and slicing through side-on swells. A stray bird or two intermittently flew close to give us the once-over. We surprised a couple of copulating turtles which instantly dived, united. I was beckoned to the bow. Porpoises were playing tag with us. Later, we sped by a large, lazily flapping ray.

Cocos grew taller and more defined. Then we were close enough to see her as she really is – wild and beautiful, wearing every shade of green. Definitely a she – a beckoning, challenging, haughty and majestic female. I was immediately attracted to her.

The boat became as a toy against her high cliffs which quickly blocked out my view of the mountain. Waterfalls, white as a detergent company's boast, tumbled and cascaded through the high greenery to the sea. Flocks of screeching boobies, frigate birds and terns escorted us across the mouth of a very large bay which presented such a backdrop that I fully expected to see a privateer riding at anchor.

'*Bahia* Chatham,' someone said. It was here, according to one map I possessed, that an Englishman, Captain Thompson of the *Mary Dier*, hid the loot of Lima Cathedral in 1821 after slitting the throats of Spanish guards and throwing their bodies to the sharks.

I saw that Chatham's adjoining bay, Weston, was little more than a shallow arc. Its white rock cliffs were honeycombed at the waterline, and it was in one of those yawning caves that the corsair, Benito Bonito,

16

had supposedly hidden silver ingots and pieces of eight. A century later, two desperate and foolhardy men dived into one of those black holes to find the treasure. They were never seen again. Cocos's waters have always been known as the breeding ground of many kinds of shark, including the great white and hideous hammerhead. A British Admiralty information sheet confirmed this and explained why Cocos had never been earmarked as a tourist attraction: 'There is generally a heavy surf all round the island. Mosquitoes and small red ants are pests. Fish are plentiful, but must be landed smartly as the sharks, which are numerous, attempt to appropriate the hooked meat. Owing to its rugged, mountainous nature, thick undergrowth, the lack of bathing beaches, and the prevalence of sharks, the island does not offer any recreational facilities.'

The western arm of Weston Bay was long and narrow. Formerly, it was called Morgan Point after the pirate who, like the rest of his kind, often pulled in to taste the 'sweet waters of Cocos' and eat the coconuts. When the island was later used by Costa Rica as a penal settlement for illegal tree-cutters and maltreaters of Indians, the name was changed to Penitentiary Point.

As we slowly passed this high promontory, the greenery gradually divided to reveal yet another bay, narrower than the other two, deep-set and walled-in by lofty, densely-covered, steep slopes. At the far end I could see a long row of coconut palms behind a strip of sandy beach where white surf pounded, and I was instantly reminded of Robert Louis Stevenson's *Treasure Island*.

We turned into this bay, which bore the name of Lionel Wafer, pirate surgeon and sailing companion of William Dampier. A very high waterfall cascaded on our right. On our left I saw that a natural tunnel pierced the promontory, joining the two bays.

There was no sign of the resident guards' patrol boat. Everything ashore was ghostly still except for the swaying palms. Standing forward at the rail as we slowly swished our way in, I could see right down to 'foul anchor bottom' through thirty feet of water. I looked for sharks but saw only shoals of small, brightly coloured fish. The rocks beneath gave way to a swath of golden sand. On the third attempt, the anchor bit and held. Engines were stopped. Only the cries of the birds broke the silence.

Shielding my eyes against the low morning sun, I discerned a corrugated iron shed through the trees. To its right was the mouth of a wide river, with a small boat on the sandy bank. A broad band of green separated the white spume of breaking waves from the blue of the water where we lay at anchor.

The ship's small skiff was lowered, and it bobbed like a cork in the high swells sweeping into the bay. Cargo-laden, it sat very low in the water as it took me towards the distant beach. I dipped my hand. The sea felt hot.

A brown figure suddenly emerged from the trees, walked along the beach and stood in front of a particular palm, indicating a path through the submerged rocks. The skiff changed course and made straight for him. I watched the pounding surf draw closer. Then it caught us, lifting the skiff high and hurtling it towards the beach. The outboard motor was quickly cut and tilted. The bow thudded into the sand, and the propelling wave swept on over, drenching us. More brown figures appeared as we jumped into the water. They helped drag the skiff to safety. Then my hand was being pumped vigorously. My kitbag, tent, suitcase and provisions were taken and I was led up and along the beach to the shed, which materialised into two large buildings, one behind the other.

The front one housed the kitchen, open-walled dining area, food store and radio with sleeping quarters above. The rear shed resembled a quartermaster's depot. One of the 'resident' guards spoke a little English. During a welcome meal I asked him where the patrol boat was. He pointed to FP407. 'No,' I said, 'the boat used here to patrol the waters.' He translated my question to the others. They roared with laughter and were still laughing when they took me to the small craft by the river. It was a rather battered eighteen-foot, fibreglass lake boat, *sans* engine, rowlocks, oars and seats, with holes gaping in the dash-board where instruments and controls had once been.

'This, patrol boat,' said the guard, and off they all went into more peals of laughter. It was a good joke. I realised, also, that it was a clever piece of propaganda put out by the government to warn off the tuna boats. 'You no tell anyone,' said the guard, still smiling. My second question took the smile from his face.

'How long would it take to get to the mountain' He turned to the others. Exchanges of voluble Spanish. He turned back to me.

'Not possible go mountain, *Señor*. Jungle. No go far. All stay camp.'

'Balls,' I said. 'How long do you think it would take to get to the mountain?' More exchanges in Spanish and hard looks at me. 'Maybe five days. Not possible. You no go. Dangerous.'

The next thing I knew, my tent and provisions had been locked away and my kitbag and suitcase were opened to make sure I had nothing illegal. Then they were placed on a camp bed above the food store.

'This will obviously need working on,' I told myself. 'Just bide your time, Kingsland.'

Shortly afterwards, FP407 sounded her hooter, calling personnel from shore, and I discovered why there had been so many people on board. Six of them hadn't been crew at all, but the change of guard.

They revealed themselves by not getting into the skiff. I studied them. The two who carried very secondhand US Army carbines were hardly out of their teens and I recognised them as being very friendly towards me on the boat. They had introduced themselves as Carlos and Joni. Both they and a Charles-Atlas-built young man called Rafael had the kindly, brown-eyed features of the Costa Rican, while two others, Ismael and Alfredo, who were in their later twenties, reminded me of the Spanish *con-quistadores*. The sixth guard was none other than FP407's cook! Named Pedro, he was sixty-four and looked not a little like Gandhi.

Just before FP407 reached the end of the Point, she gave three long whoops on her hooter. The salute was returned by Carlos and Joni, firing several rounds into the air. The hills threw back the reports from all directions. As the boat slipped out of sight, the six guards pretended to cry.

I looked up into the bright sky. High above the tree-clad hills, frigate birds sailed and whirled on motionless wings. I let my gaze travel a full circle. On three sides we were penned in by lofty walls of lichen-covered trees, entangled vines, ferns and tall grass. The fourth side held nothing but the emptiness of the Pacific, undisturbed for as far as the eye could see. There was no escape. We were completely alone. Like prisoner and escort, we marched back up the beach, and an amusing thought occurred over my situation – 'Welcome to Alcatraz'.

It was amazing how quickly the seven of us became firm friends, exploring the beaches and Wafer's valley, pig and deer hunting and generally settling in. Carlos insisted on being my Spanish teacher by pointing to everything and naming it. Sometimes my head would reel with his persistence. But no matter how I tried, even when I was able to construct small sentences in Spanish, I could never induce them to go to the mountain. And they were very firm about one thing – I was not to go anywhere on my own.

'We have instructions to watch you and look after you,' said Carlos. I told them it didn't matter as friends of mine would probably be arriving on the next boat and they would accompany me to the mountain. 'That's OK,' said Carlos. 'Two people, yes. One person, no.'

I found Wafer Valley to be extraordinarily beautiful and on one occasion, after promising I would not go farther afield, I discovered with immense joy a glade which reminded me of rural England. It was an insect-droning day and, instead of making the usual bee-line to where our local River Genio curved back round to pass through a wild banana grove, I followed the arc of the river itself. I hacked through a thicket and there it was – sparkling water tumbling over pebbles like an English trout stream.

The banks were graced with slender trunks and fronds that sliced the sunlight into gentle dreaminess. I would often sit there when I needed to be away from the incessant Spanish and to think to myself in English.

Everywhere in the valley was luxuriance and an amazing tenacity to life that filled my body and being. Paradoxically, there was also sudden and inevitable death. Yet, on Cocos, the dead quickly became part of the living. Nothing was ever wasted. Dead leaves became fertiliser. Dead animals became food. A lifeless body would disappear overnight without trace, bones scrunched by wild pig and the platter licked clean by birds, beetles, ants and insects.

To live with Cocos, I soon realised that I had to become as she was – wild, relentless and cruel. Otherwise I would be doomed, and too dismal to recognise her natural beauty. I quickly adjusted to the fact that I had to kill to live, to be dispassionate when slitting an animal's throat, even a baby deer's.

Two

IT WAS, undoubtedly, our first shark fishing expedition in that old boat, which, I learned, had been impounded from an illegal trawler, that created a strong bond of friendship between the guards and me which was further cemented by a secondary incident which almost cost us our lives.

After two weeks of exasperating fishing from the shore with its obstacles of surf and hook-snaring rocks – and catching only small fish in the river mouth with the incoming tides – I suggested we put to sea. We made four wooden paddles. For an anchor we used a hundred feet of rope with a hardened sack of cement tied on the end.

At first, only Carlos, Joni, Rafael and I were going. We had made the paddles. Just as we were taking our four corner positions on the four-inch-wide top of the tumblehome, Ismael and Alfredo joined us with spades and sat opposite each other in the centre. Pedro strolled down to wish us *bon voyage*. He was still waving half an hour later when we finally cleared the surf! My back muscles were by then painful knots of rigidity. We decided to anchor in the western section of the bay near a small rock. As Rafael threw the anchor over, boobies swooped close to our heads. We beat them away with paddles and spades. The bait bucket had been sent flying in the surf and hermit crabs were crawling everywhere, nipping our bare feet. I contemplated our fishing tackle – hand lines wound round rough pieces of wood, an assortment of rusty fish hooks, and weights made of nuts, washers, screws and nails. Carlos had made a gaff with a bar of rusty iron and a large fishing hook spliced on with wire. An almond branch, gnarled at one end, was to serve as a cudgel.

For half an hour we did nothing but feed hermit crabs to ugly black 'chancho' (pig) fish, so adept were they at picking the hook clean. When we were almost out of crabs, Joni and I caught one each. They were about nine inches long and had a single hinged barb on their backs. Scavengers, they were not recommended for eating. Cutting slices off them for bait, we caught six more and decided to move on to another spot. Off the high waterfall, Rafael again threw out our bag of cement anchor. It almost tore the bow off when the rope went taut. Obviously, we were in more than a hundred feet of water!

We baited up with chancho fish and played out our lines to the full. Then we lolled in the bottom of the boat under the tropical sun, eating a picnic of hard biscuits, margarine, tinned meat and natural fruit juice. Alfredo, who had been terribly seasick on FP407, promptly brought his up over the side to the accompaniment of the others' shouts of glee. He crawled under the half-decking and lay there. Our shoulders were beginning to burn, so we donned lightweight shirts. Carlos continued with my Spanish lesson while the others talked among themselves.

Soon, we were all in a drowsy state and sprawled ourselves out after checking our lines. Some time later, I poked my head, dreamily, over the side. The island seemed so far away. My God! It was! We had drifted, bag of cement and all!

Immediate action stations. Important things first, we all had a piss over the side. Then Alfredo was kicked awake and hauled to his spade-paddling position. The entire island disappeared every time we dipped to the bottom of each trough. Carlos was paddling and yelling 'Push!' – the only English word he had learned from me – even before Rafael and I had got the bag of cement into the boat.

Joni said something which I interpreted as 'Costa Rica next stop' which brought laughter all round. Even the ashen-faced Alfredo managed a smile. Strangely enough I felt completely safe and at home with those boys. They certainly knew about boats. 'Probably born on one,' I thought. I also had confidence in that old tub. She had a lot of float and wouldn't sink easily.

Our arms were numb, mechanical instruments when we reached Cascara Rock, two hundred yards off Penitentiary Point, and out of the drag of Cocos's constant two-knot current. Every ten minutes we had changed positions to give tortured muscles a rest. We stood, holding the sides of that tossing, rolling boat, stretching with relief and a feeling that we had achieved something. Carlos asked me if I still wanted to fish. 'Yes,' I said, and Alfredo immediately retched over the side. Ismael made him a pillow of his tunic and empty rucksack. He took it and crawled back under the half-decking.

We paddled in close to the rock until it seemed we would be dashed to pieces. The bag of cement went over and hit bottom seventy-five feet below. We were attaching bait when Joni began whooping and pointing with excitement. A black fin was cutting the water towards us. Another fin appeared. Then two more. Then another. We were surrounded! There were only three lines with leaders of fencing wire attached and it was agreed that Carlos, Joni and Rafael would use them. I elected myself gaffman and Ismael was handed the cudgel. Rafael was first to bait his hook. Three fins appeared together between the boat and the rock. Rafael hurled his line straight at them. The instant bite almost took him overboard. Then the line was burning his hands. We all rushed to help and the boat went over and water gushed over the side. Loud screams. I grabbed Carlos to get my balance, he grabbed Joni and Joni clutched at Rafael. Only Rafael's strength stopped us from joining the sharks. The Spanish that poured forth was probably disgusting as we scrambled in panic to space ourselves and right the boat.

Naturally, Rafael had let go of the line. It was speeding over the side, its wooden spindle spinning crazily and trying to leap overboard. Carlos got both hands on it just as it reached the end of its length. He was almost jerked off his feet, but he held on.

I helped him to haul in. Sometimes the weight on the end was like two sacks of cement. When the line was vertical in the water it suddenly went slack. Had we lost it? We hauled in rapidly. No. It was still there. Coming up fast! We pulled harder and faster to keep the line taut.

The shark shot out of the water, jaws gaping. Its top rows of curved teeth clunked hard on the rim of the boat, and there it hung, studying us with its hard eyes and tilting the boat lower and lower. We leapt to the other side as counter-balance and, before anyone could stop him, Ismael stepped in and fetched the shark a mighty thwack between the eyes with the cudgel. It went completely berserk, sending water all over us and into the boat with its threshing tail. Then it arched its back like a huge bow, snapped straight with a jerk that made us all clutch the sides for support, and leapt free of the boat, snapping the hook at the same time. Ismael recieved a sound ticking off for hitting the shark so soon.

Only Carlos and Joni fished as Rafael and I sorted out his tangled line and wound it on to its piece of wood. Joni stood by the half-decking, Carlos at the stern. Not a bite. Obviously, the sharks had gone. Carlos and I were determined to bring them back. We picked up the last two chancho fish, held them over the side and cut them in half. Blood and guts poured into the water. Both lines were brought in and re-baited with half a fish. 'Now we catch giant shark, Mr Gerald,' he said, as he took hold of his line again.

God only knew what monster would come sniffing round with all that blood in the water. I sat next to Joni. We smoked a cigarette and 'discussed' how lovely the island looked from there.

He was smiling at my frustration as I tried to tell him that I thought Cocos was a beautiful lady in green. Then he wasn't smiling. He began staggering with outstretched arms towards the stern. Both hands clutched the taut line and he looked back at me, imploringly. Just as he was about to step over the stern in an attempt to be a second Jesus Christ, Carlos grabbed him. This checked the line, but they couldn't pull it in. Whatever was down there wasn't budging. Rafael took up position in the centre of the boat, grasped the line behind them and pulled. The thing below began to give. Steadily, with all three pulling in unison, the line came in. I picked up the gaff.

This shark behaved differently from the other one. It allowed itself to be pulled up with intermittent spasms of resistance which bent the three of them almost double. Then we could see its shape through the water. It was big. When it surfaced five yards away it was huge. Incredibly, it behaved like a dog on a lead. Tug the line and it would follow. It appeared completely calm and oblivious to us. Now and then it made a spitting motion.

Carlos pulled it in close alongside and I could see that it was slightly longer than half the length of the boat. Carlos and Joni now had the leader in their hands. Impatiently, they tried to lift the shark's head out of the water before I could put the gaff in. As soon as one gill broke surface, the shark went crazy, slamming the boat and drenching us. Gill back under the water and it was docile again. I struck with the gaff. It bounced off! More threshing. I told Carlos to lead it right in close behind the boat then we wouldn't tip over. He did so, whistling and cooing to it. I brought the gaff down with both hands like swinging an axe. The point went in about a foot behind the dorsal fin and held on a two-inch strip of skin. The frenzied jerkings ran up the gaff, along my arms and thumped my arm sockets. I held on grimly, using the rim of the boat as a fulcrum and pushing down with all my weight. The shark's flailing tail came out of the water, but, try as they might, Carlos and Joni couldn't lift the front end. I could see blood on their hands. The rusty leader was cutting them. Rafael took over from Joni. Soon, his hands were bleeding, too. Ismael came up with the answer by twisting the leader round the cudgel. Now they had handles to lift with. 'Push!' yelled Carlos. The stupid thought came to me that I must teach him the word 'pull'. But we all knew what he meant. Gradually we got that fighting, twisting shark up on to the stern. The worst part was to come. That was when it landed with a juddering thud into the bottom of the boat.

Twelve feet by six feet doesn't allow five people much

room with a nine-foot shark that has only one intense aim in life – to bite the legs off those five people. With its nineteen-inch jaws revealing twelve rows of teeth it made for Carlos. He was nearest. But not for long. He leapt high into the air, dashed past me and jumped on to the half-decking. Joni and Rafael were perched miraculously on the rim. Ismael wrenched the cudgel free, swung hard, missed and almost knocked a hole clean through the bottom. Screeches.

Not having legs didn't seem to bother the shark over-much. It snaked and weaved like a crocodile. I held on to the gaff like I had once held an enraged shorthorn bull on a pole. Ismael was leaping about like a Morris dancer. His second blow caught the shark across the back of the neck. The rusty gaff was torn from my hands, taking a piece of skin with it. The gaff began knocking hell out of the boat. I managed to get hold of it again. If only that bloody boat would have kept still. Not only was I trying to hold the shark, I was holding on to the gaff for dear life to stop me from going overboard! Carlos joined me. Bumping shoulder to shoulder as we stood either side of the shark, we managed to lift the tail, keeping the head reasonably still. Ismael took aim. Wham! A direct hit on the shark's head. Encouraging yells. Wham! Wham! Wham! Three more palpable hits. The shark was definitely slowing. Blood began seeping through its gills.

Ismael swung the cudgel with extreme intensity. The blood began to pour forth from the gills and over the vicious teeth. Two more whams and it stopped moving. We pronounced it dead. We stood there panting and exhausted. Then Carlos's face took on the expression of the proud hunter and he gave the shark a kick. The tail convulsed and thudded against the side. Carlos sprang back amid hoots of laughter. Rafael playfully jabbed him up the arse with his fingers and shouted 'boo' in Spanish. Carlos almost jumped overboard. Then we were all roaring with laughter. They clapped me on the back, gripped me by the hand and declared me truly *amigo*.

Blood-red water swirled over the bottom of the boat,

washing an inert Alfredo. He was kicked awake again. An hour later, when we entered Wafer's surf, the sun had dipped below the bay's west ridge. We went through the breakers, pretending we were in *Hawaii Five-O*, or *Hawaii Cinco-Zero* as they knew it, their whoops reverberating round the hills.

After supper, we sharpened our knives and walked along the path to the boat. There was no cloud and the equatorial stars festooned the entire sky in bunches and singles. The Milky Way was exceptionally aglow. I looked northwards and recognised the broken-handled saucepan – a most irregular shape from there – and I wondered what my three sons were doing.

The tide was almost in, spent waves rippling up the river. We removed the shark by tipping the boat and rolling it out. Tying a rope behind the large tail fin, we dragged the shark up the bank, then beached the boat. The shark had been dead for at least two hours, yet suddenly its body twitched and its entire stomach spewed out of its mouth, pink and about four feet long. The hook was embedded in the gut wall, leader trailing. We had cut off the line. There was much excitement over the ejected stomach.

None of us had skinned a shark. No one knew where to start.

I suggested we cut it in two just behind the arse. The tail was all meat and would give us experience of skinning before tackling the ponderous body with its guts inside. Carlos drew his knife across the shark's back. Not a mark. He used the knife as a saw. Still no mark. He felt the blade. Blunt. He re-sharpened it on the sandpaper skin.

I made another suggestion – stab the point in first. Carlos drove down hard. The blade went in a couple of inches. The shark flattened him with its tail. We picked him up. He told me to have a go. I decided on a different tactic. I stood by the shark's wide back near the gills, facing towards the head, and told everyone to pull hard on the rope to keep the tail still. My knife was a handsome six-inch blade stiletto which I had bought in London and was much coveted by the guards. I stabbed hard into the back of the shark's neck. The blade went in about four inches. The shark twisted its body into an S-shape, snapped straight and knocked me over. What chance, I thought, would a person have against such a creature in its own domain, the sea? It was acquitting itself well in ours, and it was supposed to be dead.

There were now two knives sticking out of the body. I grasped the handle of mine, sawed up and down and moved with the swinging head. Every so often, the shark's eyeballs rolled alarmingly. Four of us took turns to cut through the neck and resharpen our knives on the skin. When the neck cartilage was exposed, Joni used the machete. As the blade went through, the body convulsed into a muscle-taut S again, then relaxed and slowly straightened. It took us two hours to skin and cut up that shark. When any one of us discovered a better way of doing it, he told the others. We were quite adept by the time we finished. The flesh was cut into strips for pickling and smoking. We had enough shark meat to last us for nearly a week. I noticed that even at that late stage after death, if I touched a piece of flesh it would twitch.

With all the subsequent experiences I had with sharks, I soon found that the most effective way of killing them was to knife them in the back of the neck just behind the flat of their heads. Their tenacity for being like Dracula's 'un-dead' really astounded me at times. There was one occasion when my three sons, then aged fifteen, fourteen and twelve were living with me on the island in Chatham Bay, that I discovered that a belly thrust with a knife was completely ineffectual. The boys' fishing from the mid-rock promontory, which divides the bay in two, was suddenly spoilt by a six-foot great white which kept taking their lines. After three unsuccessful attempts we finally landed it after it had raced full-tilt into a rock and stunned itself. We hauled it on to the rock and I immediately slit it open, eviscerated it and threw the offal into the water. With the hook still inside it, the shark convulsed and slithered into the sea. To our amazement it not only began swimming but snapped at and swallowed pieces of its own insides, which instantly came back out through the long belly slit. Slightly nauseated, we pulled it back on to the rock and I put the knife in behind the head.

In all the ten months I was on Cocos Island, I never saw a shark attack a boat or a raft. The boys and I made an extremely good one of balsa wood which manoeuvred well and, because it was formed with rope ties instead of nails, gave to the movement of waves and was not dashed to pieces by the surf.

One sunny afternoon we went right out to the point of the bay's western arm where I knew the hammerheads to be. In the sheltered, pond-still water, we submerged our faces and saw twenty-one hammerheads lazily swimming near the bottom. Two came up to look at us and I judged them to be twelve and fifteen feet in length. We jerked our faces from the water and the sharks came up so close that we could see their nostrils as well as their beady eyes on the wide head extensions which made them look so hideous.

'It's all right,' I soothed my sons, 'they won't hurt you.' To prove my words the two hammerheads swam down to rejoin their monstrous mates.

On another occasion, our raft was suddenly surrounded by a shoal of baby blue sharks. 'Get your arse out of the water, Redmond,' I ordered my youngest son. It was a frequent command. He was smaller than his brothers and his backside kept slipping down between the spars. The baby blues attacked the bait ferociously. Unfortunately, as soon as one got hooked, it was devoured by its brothers and sisters. We tricked them by lowering only one line. When the shoal was concentrating on the hooked victim, a second line was lowered to hook a peripheral straggler and we were able to land it whole.

'I think their mum's arrived,' said Roddick, the eldest, quietly. I looked down, and under us was a magnificent adult blue shark, and I could not but help admire its streamlined beauty.

Chatham Bay's largest resident was a manta ray, much,

much longer and wider than our fifteen-foot by eight-foot raft. The boys named it 'Charlie' and we would often encounter it. Their first sighting of this awesome creature was on the raft's test run.

Rory, the middle son, announced its first appearance with a loud 'Eek! What's that, Dad?' Two large black fins, some eighteen feet apart, were dipping in and out of the water as they approached. I had seen it often when out in the guards' boat.

'Just sit quietly,' I said. The manta ray was definitely intent on investigating us. The boys' faces filled with wonderment, awe and fright. We could see its wide-set eyes and mouth like a whale's.

With his usual pessimistic wit, Roddick said: 'Plopsy (his favourite word for me), I hope it doesn't think this raft is a female. I can just see the headline: "Four intrepid warriors fucked to death by manta ray".'

As the ray passed beneath us, its fins breaking surface on either side, Redmond whispered: 'Do you think it is going to give us a piggy-back to shore?' Then, when its long, pointed tail had cleared us, he said, smugly: 'Roddick used the F word, Dad.'

'Stop being a grass, Redmond,' said Rory. 'Make him get his arse out of the water, Dad. It causes drag.'

Of all the shark meat, that of the two-foot-long babies was the best. Deep fried it has the taste of plaice. And baby blue was decidedly the favourite. In the larger sharks, the meat could either taste like cod or be bland, readily taking on any flavour mixed with it. That only when it was of a dry texture. When it was moist it had a strong, pungent flavour.

There were many species in the two bays, but generally I found the blue shark appeared to frequent Chatham. In Wafer, the guards and I, during the first six months I was with them, caught white-tip and black-tip. For some unknown reason, it was I who always seemed to catch the sharks, mostly between three and six feet in length – they simply went to my hook first. Naturally, with each monthly change of guard I was always introduced by the retiring guard as 'Mr Gerald, shark man' and the life-pattern with each new set of guards was the same – I would take them out in the boat shark fishing. And always there was initial alarm on the new guards' faces when I pulled the first twisting, fighting shark into the boat, and subsequent relief when I knifed it.

That first shark the guards and I caught was a thresher shark, related to the great white but slimmer, its long tail fin adding more to its length. From what I saw of great whites, later, with their deeper chests and bellies, I don't think we would have got a nine-footer into the boat with the Heath Robinson tackle we had.

But all sharks, from four feet upwards, are formidable creatures. Even a four-footer can rip a man's calf completely away.

Three

A FEW DAYS before the initial shark fishing incident, something had occurred which partially estranged me from the guards, and I was pleased that the fishing trip re-cemented and strengthened our friendship. The previous guards had left behind a dog called Pinto which Carlos claimed as his. It resembled a lurcher and had an independent hunting spirit. Whenever it bayed a pig, the guards never fired from a distance but always crept up and put the muzzle in the mesmerised pig's ear before pulling the trigger. Also, when firing at a deer from an un-missable distance the bullets went far and wide. The only time they could hit anything was at point-blank range. The only time they had shot a deer in that first week was when Pinto had caused it to run blindly almost into the end of the barrel.

I told them that I knew about US .30 carbines as I had often swapped my .303 Lee Enfield with an American's carbine in Korea, and I suggested we had some target practice. The target was the end of an empty forty-five gallon petrol drum, in the centre of which I stuck a small white envelope. We drew back seventy-five paces and I told them to fire. They did so. There wasn't a mark on the drum. Not only had they not elevated the rear peep-hole sight, but they had looked over it, not through it. I took Joni's rifle. The back sight was filthy. I poked dried earth out of the peep-hole, pulled up the very stiff leaf sight and set the peep-hole at the one hundred mark. They all watched with great interest. I aimed low and fired. The bullet clipped the top centre of the envelope, punching a neat hole through the thick metal. They were very excited and wanted to know what I had done. I was forced to use miming actions because my Spanish was still negligible.

Rifle oil, brushes and cloth were produced and the sights vigorously cleaned.

Then they began firing. The drum was riddled. Joni and Ismael put holes almost dead centre of the envelope. They were the only two who showed real interest in shooting and subsequently bagged several deer.

I noticed that the guards were suddenly rather wary and shy of me. Several times I caught Carlos looking at me strangely. The questions came that evening at dinner. What was war like? Was I frightened? What was fighting like? I was able to answer them with the aid of a small Spanish–English dictionary I had with me. Carlos asked the inevitable 'How many men did you kill?' I hesitated – then told them a half-truth. 'Five,' I said. It sounded shocking to me. I know it was to them, so how could I possibly tell them that I had caused hundreds of deaths by ordering and directing shell-fire?

'With rifle?' Carlos asked. 'With rifle,' I replied.

They were intensely interested in war, yet hated the thought of it. So did I. It had left an impression on me that would never leave.

They talked among themselves afterwards. Next day it appeared they were avoiding me. On one occasion, Carlos looked me straight in the eyes across the dining table and said, flatly: *'Cinco hombres,'* and shook his head in condemnation. I felt I was on trial for murder and an awful sadness and loneliness engulfed me. Then their camaraderie returned. It was as if they had all decided to forgive me.

A week later I was able to show them more of what a rifle could do, and I learned what a lethal weapon the machete was.

I longed to see some other part of Cocos than Wafer valley, and I couldn't be blamed for feeling somewhat a prisoner. I produced my large map of the island, specially prepared that year, 1978, for the Costa Rican government from aerial photographs, and spread it on the dining table during mid-morning coffee.

Pointing to Chatham I said: 'According to my papers,

34

the crew of a whaleship crossed the ridge from Chatham to here in 1838. Why don't we have a try, walking from here to Chatham?'

'How far is it?' asked Carlos. They all crowded round as I measured the distance with a ruler. 'About one and a quarter miles as the frigatebird flies,' I enthused. 'Come on, we can keep the sea in sight for most of the way so that we don't get lost.'

To my joy, they all wanted to go. 'OK,' said Carlos. '*Mañana.*'

In the morning, Pedro decided he was too old for the journey. 'He's antique,' said Carlos. 'But I am antique, too,' I protested.

Carlos flung his arms around me with a wild whoop. 'You are not antique, you are *el tigre*,' he cried, and he pranced and growled like a tiger. Then he paid me a high compliment – he handed me a rifle. 'We take machetes,' he said, 'you do the shooting.' As we picked our way through the swamp at the rear of the camp to where the steep slope of the seven-hundred-foot ridge began, I felt not a little like a white safari hunter of old, with his coolies cutting the trail before him.

Every inch of the way for the first three hundred feet up from the valley floor had to be laboriously cut. Then we were on rain forest slopes where the occlusion of light prevented much undergrowth, and we were able to climb fairly easily. Sweat poured from us, and Carlos showed me how to pull down the blades of the lower clinging epiphytes so that the pure rain water held in their bases ran into my mouth.

It took us two and a half hours to reach the top and there we rested, ate a small picnic and admired the view of Wafer and Weston bays. Then we set off briskly along the fairly level floor of the rain forest, marking the trees for our return. Twice we were forced to go down into small valleys and cut through thick undergrowth. We made good progress until we came upon a dense and very tough wild coffee plantation which slowed our progress considerably. When we emerged, we found we were out of the rain forest and confronted by a solid barrier of ferns that towered above us for about thirty yards.

The ferns gave way to shoulder-high grass, dotted with very tall bush clumps and lone trees. Now and then we heard the thunder of hooves and cracking undergrowth, heralded by Pinto's barking, first in one direction, then another. Hooves were suddenly galloping towards us. I eased off the safety-catch. A large deer leapt into the cut path amongst us, almost flooring Carlos. Surprise and fright caused it to pause. In that split-second Carlos's machete flashed, laying the deer's flesh open to the bone from hip to knee. A mighty leap and the deer disappeared into the thick grass with everyone screaming at me why hadn't I shot it.

'I want to show you something,' I said, turned and fired at a tree a few yards away. Its trunk was about two feet six inches thick. 'Now look on the other side,' I told them. They were amazed to find a neat hole where the bullet came out. 'If you ever get to the Nicaraguan Frontier,' I went on, 'don't hide behind trees. The bullet will go through you as well. What you see in films is a load of rubbish, especially hiding behind doors and upturned tables.'

It probably took me five minutes to say all that, but they were suitably impressed. I was quite impressed, too – with the speed of Carlos's machete. I was also upset by the cruelty of the ineffectual wound it had inflicted. We continued and felt that the ground was descending. A shout went up from the leaders. Before us was a plain of thigh-high grass and a magnificent view. We were on Chatham's famous Green Hill, referred to by several treasure seekers and called a 'bastion' by one who tried to climb it in 1891.

From our vantage point, eight hundred feet high, we could see the entire bay – except for the central shore immediately below – with Cocos's largest satellite rock island, Manuelita, sitting off the western arm. The distant, misty horizon was a continuous sweep for almost a hundred and eighty degrees. Surely, I thought, a ship must be visible somewhere in that vast arc. But there was nothing.

We split into three sets of two to get down. I was paired with Rafael. Our choice was a deep gully, filled with ferns taller than we were, which ended, almost to our cost, on the edge of a hundred-foot cliff. We clambered up the high bank to our left and found we were about a third of the way down. On a nearby dead tree, unflinching frigatebirds regarded us with solemn curiosity. They looked like vultures, waiting for a meal to drop by – us. We were both exhausted, baking hot, thirsty and in agony with ant bites. Rafael had been bitten on the balls. I felt infinitely sorry for him as he dropped his shorts to inspect the damage.

A reasonable enough slope to our left showed promise if we descended across it. We slipped and slid for most of the way. The sea was still quite a way below when both of my feet shot from under me and I rolled and slithered for yards before coming to rest in gorse.

It was as I sat, picking thorns from my tortured flesh and poking mud out of the rifle barrel with a piece of stiff grass, that I called that hill a much stronger word than bastion. Rafael wholeheartedly agreed. Chuckling, he went on down. Suddenly, he shouted: 'Look!'

There, in a lagoon of turquoise, was a wondrous sight – a stag and doe swimming side by side as though in some fabulous Disney scene. I watched, spellbound. 'Shoot! Shoot!' yelled Rafael. He began running, brandishing his machete, slipped and disappeared.

But I couldn't shoot – not because of any traces of mud in the barrel, the bullet would blow those out – it was just too beautiful, too tranquil a sight. Now and then the stag turned its majestically antlered head to make sure his mate was all right. I saw the reason for his concern. Pinto was nosing through the water after them. Shouts reached me and I saw Carlos and Joni running along the beach, their machetes pointing the way like lances. The stag and doe passed from my view behind an intervening large tree. I decided to get down there fast.

I crashed into the tree, slithered round the trunk and went on down through small bushes and gorse. The next minutes were a blurred series of falls and flashes of the scene below. The stag was first out of the sea, and galloped, with the doe close behind, to a far corner of the beach where it was bordered in front by a high bank of grass, and by high rock on the side. The stag cleared the bank to disappear into the undergrowth. The doe didn't quite make it and slithered back, her front hooves flailing the bank. She was trapped but I couldn't shoot because Carlos and Joni were directly in my line of fire. I floundered on down through grass and ferns.

There was only one possible way of escape for the doe – the sea. She bucked down the beach with short jumps. Pinto met her at the water's edge. She reared and veered from the dog. Pinto was round to the other side like a flash, leapt, and his teeth held on her neck, pulling her to her knees. She regained her feet, swung and loosened the dog's grip. Trying to balance I raised the rifle. Carlos and Joni came into my sights. 'Fuck it!' I said, and stumbled on down. As I went arse over tit into a clump of ferns, the machetes were whirling. I was hanging on broken ferns above an eight-foot drop to the beach. I let go and thudded into the sand. But I was too late. The doe had died horribly. Bone gleamed white through numerous back cuts, entrails bulged through belly slashes; one leg was severed through the bone and hanging on flesh, one ear was missing and an eyeball hung down a lacerated face.

'*Bueno*, eh, Mr Gerald?' asked a triumphant Carlos, as I panted to them. '*Bueno*, eh?'

I was seething with soreness, and over the way the doe had been killed. 'No, it's not bloody *bueno*,' I screamed at them in English. 'You're a cruel bunch of bastards and you ought to be shot.'

They didn't understand the words. They just didn't understand. They looked at me in embarrassed, hurt and puzzled silence. 'What the hell?' I thought. 'Who am I to criticise? Their ways are their ways.'

I realised they were looking to me for praise. 'I'm sorry,' I said in Spanish. 'Yes, Carlos, it's *bueno*. Truly *bueno*.'

Rafael, then Ismael and Alfredo, arrived and said as much, too.

'Why didn't you shoot?' Rafael accused. I explained as best I could. Carlos threw an arm round my shoulders. 'Yes,' he said, hugging me, 'I understand. *Amigos*?' He offered his hand. '*Amigos*,' I said, and shook it.

Pinto was getting his share of blood from the weals. We examined him. Miraculously, he had been untouched by the blades. Four months later, in the November when Carlos returned to Cocos, Pinto was not to be so fortunate.

Rows had been constantly flaring between Carlos and a guard called Vargas, who featured quite a lot in my life on the island. After one particularly heated argument, Carlos picked up a machete, muttered something about Vargas's early demise, and stormed off along the beach with Pinto. I sat for a minute or two, then decided to go after him. He entered and went up the short trail we had made to the low ridge at the beach end of Penitentiary Point, and he disappeared behind a series of boulders. When I was three-quarters of the way across the beach, Pinto began barking, a pig began squealing and Carlos began screeching like a maniac. I broke into a run, scrambled up the trail and reached the boulders as the barks, growls, squeals and shouts became a terrifying cacophony. I dashed between the boulders and stopped. A half-grown sow was on the ground, her body a mass of red-wide cuts and blood squirting from a lethal gash in her throat. Pinto was attacking her, the blade swinging close to him, and I could see he had already been sliced. His right thigh was open for its entire length and bone glistened in a cut above his left front paw. Still the blade whirled, driven by Carlos's frustrated rage.

'Carlos!' I yelled. 'Stop it! You'll kill Pinto!' As I spoke, the blade took off the pig's right ear and laid Pinto's head open above the right eye, causing the skin to drop, closing it.

'For Christ's sake, Carlos!' I stormed, and dashed in. He realised then that I was there. His shoulders sagged and he stood limply, body soaked with sweat and his chest and belly heaving. Pinto withdrew, dazed and whimpering, his front paw raised pathetically as he seemed to be asking me a question with one brown eye. I took the machete from Carlos's unresisting hand.

I knelt to examine the dog and it was then that I fully realised how much I had changed. Death and injury from the machete didn't horrify me anymore. There was nothing new or dramatic in that. Anyone can become inured to anything. And there was no such thing as heroics and cowardice. Korea had shown me that. Usually, a frame of mind could depend on whether one had had a good shit that morning. Could bowels be the subtle difference between heroes and cowards?

I felt no emotion for the pig or the dog. 'Is he all right?' Carlos asked, quietly, when I had finished my examination. 'He'll live,' I said, bluntly, 'but he'll need some looking after.'

Carlos gathered Pinto up into his arms and carried him, cradle fashion, often kissing the dog's head. I hoisted the dead sow on to my shoulder and followed. Knowing Vargas was bound to say something inflammatory, I looked at him hard and he said nothing. All three cuts needed stitches but there were no suture needles. There was a roll of sticky bandage so I decided to do what a woman in Italy had done when my son Rory laid his knee open with a fagging hook: pinch the skin together and stick it with the bandage.

I concentrated only on the eye, allowing him to lick the other two wounds. The eyeball was intact but the bone above was chipped. Pinto was a good patient and allowed me to change his eye dressing daily. There was, for the first three days, superficial pus but, underneath, the flesh was joining healthily.

Carlos fed him the best of everything and served him condensed milk in warm water three times a day. The wounds became narrower, shallower and shorter. After five days we had to tie him up so that he wouldn't limp and whimper after us when we went hunting.

Four

IT WAS during my fifth week on the island that we embarked on the boat trip which nearly killed us. After the catching of that first big shark, there were suddenly shoals and shoals of edible fish in the bay, including the succulent Australian big-eye. Almost every day we were launching the boat to fish for them. But with the fish came the sharks and they were a decided nuisance. The story was always the same: catch one edible fish and 'Mr Nosy' would come sniffing around and frighten the shoal away.

'Carlos, my son,' I said, during lunch, 'I know what we should do this afternoon. Try fishing in Chatham.'

'Mr Gerald, are you a masochist?' he asked.

'I don't think so,' I replied. 'I can't stand pain. Why?'

'You want to carry fish up Green Hill and bring it one and a quarter miles over that ridge? If you remember, we had to cut off the legs of the deer and carry only those.'

'No!' I said. 'I don't mean walking there – I mean going by boat. There's no wind and no sign of rain. We just hug the coast.'

He asked the others what they thought. As one, they said 'yes' – even Pedro declared he was going. So it was that seven men – to say nothing of the dog – put to sea. We made fairly good time until we reached the Manuelita Strait. For thirty minutes we fought the shallows which churned the sea like rapids. Halfway through, we all exclaimed in utter astonishment. Riding gently at anchor in the centre of the bay was a three-masted brigantine. 'We're in a time warp,' I thought, 'and we are back in the days of Morgan and Dampier.' The craft was all black and trim, her masts and yardarms with folded white canvas rising in perfect harmony against the wild hills behind. She looked completely at home in those waters.

'Push!' we yelled in unison and excitement. Clear of the strait, the guards took their official tunics from under the half-decking and put them on. Then, with Pedro standing like a captain on his bridge and holding Pinto – who sat with a rope round his neck on the bow like a figurehead – we made our painfully slow but dignified advance on the 'pirates'. Cocos Island Sea Patrol was on official business – intercept and inspect the ship's documents.

Decorum and dignity slipped slightly when we observed that half the personnel lining the rail to watch the expert synchronisation of our paddles and spades, were rather fetching young ladies.

Words of exclamation, which roughly translated meant 'crumpet', brought a faster tempo and we actually produced a bow wave. The faces looking down upon us as we manoeuvred alongside bore expressions of amusement and sympathetic incredulity. The absence of trousers below tunics, suggesting half-nudity, brought forth a few giggles.

I addressed a kindly, grey-bearded, Quaker-like countenance in almost apologetic tones: 'Are you the captain?'

'I am, sir,' he replied in a strong American accent. 'What service can I be to you?'

'As you can see by their tunics,' I said, 'these boys are Costa Rican maritime police – I live with them round in Wafer Bay. They want to see your ship's papers and would like to come aboard.'

'Delighted,' said the captain, and a rope ladder was slung down to us. We all climbed up except Pedro and Pinto. Everyone crowded round for introductions and to know what on earth we were doing there. It was such a relief to speak English again, I couldn't stop talking.

'We saw you coming through the strait,' the captain casually remarked. 'We thought you had stopped off for a spot of fishing!' He looked over the side into our boat. 'It must be hard work, pushing that thing through the water. Can't they afford an engine?'

Ship's papers were produced and a legal document in Spanish satisfied Carlos. The all-timber brigantine was owned by an exclusive American school of seamanship for students who could afford the fees. They were on their way home from Peru, had encountered heavy seas and pulled into Chatham on the previous night to give the seasick ones a rest. They were sailing next morning at six. The students were mostly female and mainly from the Deep South with accents like Scarlett O'Hara's. They thought everything was 'just fascinating, simply fascinating'.

There was no fresh meat on board and sharks had bitten off quite a few fish hooks in the bay. We had a rifle with us and I suggested we went ashore in the ship's boat and we'd try to get a pig or deer. The trail the guards had cut up Green Hill was brown and clearly defined from where we were. Those who wanted to stretch their legs went ashore with us, and we found a short but strong waterfall behind an old fishermen's hut. The students made full use of it for a 'delicious' fresh water shower. As luck would have it, Joni shot a very large deer.

'You must stay the night,' said the captain on our return. 'Everyone could do with a party, and I'm sure we can rustle up a few bottles of wine to go with the venison.'

It was quite a hearty feast. I sat next to a sweet, buxom young thing from Texas. The wine was slightly acid and extremely heady after our period of temperance. Guitars strummed afterwards, the ship was rolling gently and the party took on a dreamy aspect. I already had a hangover, so stepped over a few inert bodies and went up on deck. I took a walk round in the gentle rain and found Joni and a young lass playing catch me up the rigging. He was hanging on a rope with one hand, offering her a glass half-full of red wine with the other, and telling her he was a giant tiburon, she was a little fish and he was going to eat her.

With the warm rain streaming down her happy face she implored me to tell her what he was saying. 'If you don't tell me, I'll die. Simply die!' she cried.

'Well,' I said, 'to put it in the vernacular, he's offering you a swift half and his leg over.'

'Oh, the filthy beast!' she exclaimed, and burst into peals of laughter.

The fug below was worse after the fresh air. I just wanted to put my pounding head on a pillow. Buxom Brunette led me to an upper bunk at the far end of the dining room. I stretched out with a brief thank you and drew the curtains. I awoke, sweating, wedged against the bunk wall with a tight band round my chest. It was an arm and it felt like rigor mortis had set in. I explored in the darkness. The arm was attached to the Buxom Brunette, who was completely naked. 'Oh, my head,' she moaned. 'I'm going to die. Simply die!' My head was aching, too. But, dammit, I was British. I felt it my bounden duty to ensure that her passing was a happy one.

The pain in my head had slightly subsided when Carlos awoke me. BB had gone. I never saw her again. Ships that pass in the night, I thought; except, of course . . .! My mouth felt like all the usual descriptive proverbs. Carlos looked quite miserable, too. 'We have to go,' he said. 'They are ready to sail.'

We collected various pieces of Costa Rican police attire and bodies from the floorboards. 'Good luck,' said the captain, as I half-fell over the side. 'Nice having you aboard.'

Red-eyed, sore-headed and in rather dazed belligerence, we stood in the boat, rain water round our ankles, as Pinto was lowered on a rope. The ship's sails unfurled and billowed in the early morning's misty breeze. Only the captain waved. The others were far too busy with the ropes. I felt rather like Captain Bligh as the ship drew away and was soon sending a white bow wave along her hull.

We all felt forlorn and abandoned in that waterlogged boat, far out on Chatham's grey and broad expanse of water. Carlos was the only one who could think of anything to say at that precise, poignant moment. 'Pinto shit twice on board,' he announced, spitefully. 'I hope they all tread in it!'

We howled with laughter, and patted and made such

a fuss of the dog that he began barking in bewildered excitement. As we washed ourselves with the water inside the boat, we knew – and welcomed the fact – that we were islanders once more, I more so than they. Cocos was my elected lifestyle. For them it was simply a tour of duty. But the 'islander bug' can suddenly bite the susceptible and quickly manifests itself in feelings of territorial possession and resentment of intrusion. I could sense it in the guards and it was very strong in me. We were quite happy to see our fleeting, bitter-sweet interlude with civilisation speeding, full-sail, on a wind-borne course for America. Unfortunately, that same wind was blowing in the wrong direction for us.

Pedro was ordered to start bailing as the rest of us brought the boat round and began what was to be a marathon paddling session. Ninety long, muscle-tearing minutes later we reached the lee of Penitentiary Point in Weston Bay. I knew we were in real trouble when the wind brought black cloud and driving rain. The swells grew higher with increased frequency and we could feel as well as hear and see the thunderings of the waves against the impassive cliff walls, hurtling each wave back to meet the following one, shooting the spray high and creating turbulence throughout the bay.

We looked at the tunnel which linked the waters of Weston and Wafer bays through the width of Penitentiary Point. It was only just wider than the boat, the giant swells rushing in and sucking out of it. It would save us from having to face the strong wind on the point where the trees were bent almost double. We entered and found it too narrow to use paddles.

We pushed and pulled our way along with our hands, getting them out of the way fast as each swell sent us soaring almost to within head-crushing reach of the ceiling, then plummeting so that we felt as though we were in an express lift.

The ceiling was a myriad of wide cracks and precariously held jagged pieces of rock, some pointed and hanging like stalactites, any one of which appeared to be just waiting to be vibrated free by the boat's screeching, scraping and thudding against the walls. Some of the gravity-defying rocks were so huge that, had they fallen, all of us and the boat would have been completely obliterated. I comforted myself with the thought that they had hung there for centuries – why should they fall now? The disturbing answer was, of course: 'Sod's Law. Kingsland is below them.'

The tunnel was the subject of a disagreement between two naturalist-writers, L. J. Chubb and C. L. Collenette. The former gave its length as '100–200 yards' while the latter, who claimed to have ventured through it, said it was only thirty yards long. Measuring the tunnel on my map later, I found it was a hundred and thirty-eight yards long. Small wonder that it was with immense relief we finally emerged, but that relief was instantly dashed, as the boat almost was. The full fury of the storm hit us and we were swept high and back to the huge rocks on the tunnel's northern side, the waves shooting up them as high as a house.

We were screaming at each other in panic amid frenzied paddling. Just before we could crash, we slid away in a deep ebb only to come winging back to the accompaniment of our distressful cries. Carlos rushed to the stern. He, Rafael and Joni took most of the shock with their paddles. Even so, the stern crunched sickeningly. Carlos dashed back to his position opposite me at the bow as we slid away once more in the yawning trough. Furiously, we drove in paddles and spades to give us more momentum with the outward flow. This time, most of the returning surge slid beneath us and we were several yards off the rock when the following ebb carried us out even further.

But the headwind was too much to make directly for camp. We were driven backwards towards the open sea and to where rocks glistened in the troughs.

'*La playa! La playa!*' I shouted, desperately, and pointed to the opposite shore. They instantly saw what I meant, and we aimed diagonally out and across so that we were slightly with the wind. Fright had brought Alfredo from under the half-decking where he had slunk with awful seasickness when we were mid-way across Weston Bay. As one of us collapsed for a short rest, the seventh man would take his place, and in rotation we paddled and spaded our way across the bay's raging sea. Any observer would have seen a miniature caricature, with waves proportional to boat, of a ship of yesteryear battling the storm-tossed seas where the 'Roaring Forties' blow.

A treacherous carpet of large and small black pebbles awaited on the beach for which we were heading. But we had to land. We were too exhausted to go further. Fifty feet off, Rafael threw out the bag of cement. It dragged with the force of the waves but allowed us to go in backwards with some checking. Apart from Pedro we all jumped overboard into chest-high water and tried to keep the bow head-on to the waves. Pinto was out like a shot and swam strongly for shore. Reaching the beach he was immediately knocked flying by a breaking wave. The sea was too strong for us, too. The boat was suddenly swung sideways, knocking the three of us over on my side. We were pushed flat on to our backs on the bottom as the boat sped over us. We regained our feet, spluttering, in time to see the boat, with the other three desperately trying to hold on, crash sideways on to the beach, sending poor old Pedro arse over tit over the side. We got there in time to stop the next wave rolling the boat on top of him.

Utterly knackered, we sat or lay on our backs in the pouring rain, almost inclined to let the boat get smashed to pieces. At least, we could then all walk home. Making light conversation I said: 'We should encourage tourists to come to Cocos and we'd paint big letters on the side of the boat, saying: "Heart-attack trips to Chatham".'

'Yes,' said Carlos, managing a smile. He nodded towards Pedro. 'Antique can be captain.'

'Anyway,' I continued, 'it *was* a good party.' They all spat at that, except Joni who gave me a sly wink and smile.

After the beating the boat had taken, she deserved to be seen home safely. Carlos came up with the answer – Pedro would go on ahead and get coffee and a meal ready, Carlos and I would ride the boat, keeping her off shore while the others towed her with the anchor rope. And that was how we got the old tub back. The two-mile journey had taken us four hours and forty minutes. We vowed never to go outside the bay in the boat again, and somehow I got the feeling that I would never be king of the suggestion box.

So far as the guards and I knew, only one other ship visited Cocos all the time I was there. Many small yachts of different nationalities came in for an overnight rest and always the people on board were invited to eat a meal with us. The one ship was a tuna factory, flying both the American and Costa Rican colours. I was invited on board for dinner with the captain and chief engineer, and they asked me the question many people subsequently asked: Did I consider myself to be a drop-out? 'I don't know,' I said, truthfully, and briefly outlined the story of my life.

For a start, the world had been pretty good to me. From straw-sucking farmhand, backed by an idyllic childhood only slightly marred by the war years, I had elevated myself, literally and metaphorically, by joining the Parachute Regiment, and I would thrill to the cracking sound of the opening parachute and the ensuing, incredible silence and stillness aloft. I had served my country, and my Korean War bounty put me through college for one year in order that I could enter provincial journalism, which I dearly enjoyed, finally arriving in Fleet Street to specialise in Law Courts and House of Lords reporting.

Then I met Rosemary, a strikingly beautiful ex-public schoolgirl of twenty. She was assigned to me as a cub-reporter and we fell madly and passionately in love. So intense was our relationship that we both thought it must surely blow itself out in a few weeks. But it didn't and she was cited in my wife's divorce proceedings.

'That passion lasted for seventeen years,' I said. 'And that is a long, long time.'

'What went wrong, then?' asked the captain. 'Help yourself to another beer. There's plenty more.'

The alcohol had taken a hold, and I continued to bare my soul. 'For ten years we never had a single solitary row and she presented me with three lovely sons. I had switched from newspapers to magazines and, after a brief period as first editor of Britain's leading girlie magazine *Mayfair* – Rosemary, incidentally was the first Pat Asquith who answered people's letters, but really in those days they were written by me – I launched a highly successful sex-education magazine called *Curious* which made me a wealthy man almost overnight. Rosemary was my co-director, and that was when the rows started.'

'Never involve your wife in your business affairs, old son,' said the chief engineer. 'Fatal mistake.'

'What happened then?' asked the captain. I told them that I wasn't too sure. Probably the wealth was too sudden for me to cope with. I began drinking and five years later I found I was running a country house in the Welsh mountains, three cars and two mistresses.

'Emotionally I was utterly miserable,' I said. 'From rags to riches, as it were – and for what? I'd made it, there was no more excitement and I wasn't happy.'

'I think you'd probably fucked yourself stupid. That was your trouble,' said the chief engineer.

'Sure I'm not boring you?' I asked with a trace of resentment.

'No, of course you're not,' said the captain. 'We'll let you know if you do. We'll throw you overboard to the sharks. You're not drinking very fast, you know.'

I went on with what had suddenly become a saga. 'I was then forty-four and I hated the roll of belly fat that was threatening to misshapen my tailor-made suits.' The chief engineer made an ineffectual attempt to draw in his huge gut and we all laughed. 'A mercenary friend suggested I bayoneted a few black babies in the Congo.'

'Charming friend you have,' said the captain.

'Then Rosemary announced that we were turning our backs on sex magazines and we were going to live in Italy. There, I accepted the challenge of an overgrown and abandoned vineyard, took it on lease, flattened my belly with the work and had it back in production three years later.'

'Then what?'

'Rosemary up and left me and the boys. That was the saddest day of my life. I don't think I'll ever get over losing her. I'll never be able to find anyone to take her place.'

'Why did she leave?'

'Drink. I'd tamed the vineyard and there was nothing else to do. I remember there were fifteen hundred litres of wine in the cellars and I divided the number by three hundred and sixty-five to see what my daily intake should be. I wracked my brain for some different kind of venture and I was sitting, pissed as a fart, under an olive tree when it hit me. Be the world's second Robinson Crusoe. And here I am. What do you think?'

There was a long silence. The captain broke it. 'I think you're a bloody fool. Won't she come back?'

'No. She's found someone else.'

'Oh, have another bloody beer!' said the chief engineer. 'You might just as well get pissed. God knows when you'll get your next drink.'

Five

BLONDE, BLUE-EYED and twenty-four, farmer's daughter Anne Hughes, who was to have been my first Girl Friday, soon told me the reason why I was trying to be Robinson Crusoe. 'You're just running away,' she said. 'You don't want me, you want Rosemary. I can actually feel your resentment.'

'That may be true,' I said, 'but I really do want to live on an uninhabited tropical island. When I was sitting under the olive tree, my thoughts turned to the islands outside Kure harbour in Japan. I was being taken to Singapore on the aircraft carrier *Unicorn*. Those islands were verdant, tranquil and beautiful to me after eighteen months in the front line, day after day seeing nothing but brown soil, blasted trees, barbed wire and sandbags. Those islands appeared and appealed to me as the only peaceful places left in the world. Can't you see, a desert island is, and always will be, the escapist's dream. All right, so I am running away, trying to escape, but at least I am doing it and not just dreaming as all the other armchair adventurers do.'

Anne and I had met, indirectly, because of the Costa Rican chargé d'affaires in London. He had refused to give me permission to live on the island alone. How ironical that was, I often thought afterwards.

'It is too dangerous,' he said, 'and my government would not be prepared to take the responsibility.' It was a problem I had encountered a lot in my quest for an island. 'I suggest,' he continued, 'you get a group of people together to go with you – colonise it. We have no use for the island. We know it's there, but we don't know very much about it, other than that treasure is supposed to be buried there.'

Being the romantic soul that I am, I had an instant vision of *Mutiny on the Bounty* and Pitcairn Island. So I would be a second Mr Christian not Crusoe – the nuance was slight. As my friends and acquaintances thought I was completely mad, and they had no desire to accompany me, I placed an advertisement in the travel section of London's *Time Out* magazine: 'Couples and single people wanted to colonise uninhabited tropical island.' My phone never stopped ringing for four days, and rang continually for the rest of the week. Even after that, there were sporadic calls.

I found that in this world of fast evolution of equality among the sexes, there were now *female* heavy breathers. One promised me an idyllic life of blow jobs that would blow the nuts off the palms! Other women pleaded with me to take them away because their husbands beat them. There were drug addicts ('say, man, we could make the island a marijuana plantation'), gays, kinky couples and a prostitute who insisted that her profession was essential everywhere.

I lost count of the number of people I interviewed. To many, I knew, the idea of the venture had great romantic appeal, but the actuality – that we were really going – was too much for them and they withdrew. I ended up with what I thought were nine solid, eager, determined people. But over the weeks we became like Agatha Christie's *Ten Little Niggers*. For a start, one would-be adventurer collapsed with a mammoth epileptic fit during a camping weekend on Anne's farm. His wife went off with him in the ambulance. Next, a beautician said she'd been made the offer of a job she couldn't refuse. No, of course her husband wouldn't be going without her. A third married couple, whom I liked very much, had striven for a child for years. Suddenly, they announced they had hit the jackpot. 'It must have been that camping weekend,' they said.

We four remaining 'Little Niggers' held an urgent meeting. We decided to go ahead. Money was now no problem. I had been given a substantial advance by a publisher to write a book about it all.

It was agreed that I would pay everyone's single air fare to Costa Rica. If they didn't like the island, the return was their problem.

At the eleventh hour, the deep-sea diver of our foursome said there was an emergency situation on the oil rig and he couldn't let them down. He and his twenty-year-old girlfriend would follow a month later provided I found the island suitable and safe. He would organise the shipping of our freight, containing donated tools and equipment, including a Danarm chainsaw.

So it was that only Anne and I flew in to San José. But after three weeks in a hotel room, although we were the best of friends, and still are, we knew that while I was missing Rosemary dreadfully, she missed her fiancé from whom she'd wanted time to sort herself out.

When we were told there was room on the gun-boat only for me because of extra cargo, I told her I would pay her fiancé's fare to Costa Rica. They were both welcome to come to the island on the next boat. 'And find out if the other two are coming,' I said. Alvaro had assured us there would be room. He had granted me twelve months' permission to live on the island, with option to renew.

A few days before the boat was due to bring the first change of guards – and I realised with some sadness that I would be saying goodbye to six very good friends – Carlos told me that, according to a radio message, four English people would be on board. This alleviated my sadness over the guards' pending departure, and I was looking forward, also, to receiving letters from my sons.

I was almost shattered with disappointment when the four arrived. It was very soon apparent that the two couples couldn't get on with each other and I found myself acting as mediator. They had brought a pitiful amount of provisions and had left the freight behind. But by far the greatest blow, they had no desire to help me get to the mountain. After three weeks, I suggested they left on the next boat.

'Give us a chance, Gerald,' said Anne. 'We're not like you. We find the island a bit frightening.'

Three months later I asked her for her impressions of the island, past and present. She said that when she first saw Cocos she was too sea-sick to care very much about anything. All she wanted to do was get off the boat! But, later, she had felt a quiet wonderment at finally arriving.

Like me, Anne was convinced that Cocos was a 'female' – vulnerable but strong. 'Man could easily destroy or deface her,' she said. 'Yet she could easily destroy a man.' Also like me, Anne had a great respect for the island's interminable life source which she found to be slightly overpowering, yet not depressive.

She loved the rocks and driftwood, which she called 'smoothed-down companions'; she enjoyed the comforting chatter of the crickets at night; she felt murderous to all mosquitoes, and was certain that she was more afraid of the rats than they were of her!

Anne was the only one who helped me with the vegetable garden I started for the guards, sifting the soil with her fingers and tending the fast-shooting plants. There was a strong rapport and friendship between us. As she commented, though: 'It's a shame we both love someone else, but I am always here if you get really lonely, if you see what I mean.' We both knew that I would never take up the offer.

Because of the book I was to write, the government was quite happy to keep on feeding us all. I didn't mind Anne eating the guards' food, she had had the guts to fly off with me and had been more than prepared to go on to the island had there been room on the boat, but I deeply resented the other three who just sat there, eating and doing nothing.

With the exception of Anne, to whom I could talk and relate, the others were no longer part of my book project or my story, and I told them so, flatly, adding that I deeply regretted the money I had wasted on their air fares.

Gradually, I found myself preferring the company of the guards, acting as guide to take them to Chatham or showing them the hunting trails along the valley. None of the August and September guards wanted to be on the island and often did nothing but sit, staring moodily out to sea, hoping that the boat would arrive prematurely.

Always, during the week the relief boat was expected, they would cease all hunting activities and I would go off alone with the rifle, staggering back with pig or deer on my shoulders. And I would control my anger as I watched my once would-be adventurous companions tucking into the flesh.

With the sullen, sloth-like behaviour of most of the guards during those two months, I often reflected how my position had changed since the day I had stood on the beach with Carlos and the others on the day of my arrival. I was now the one with the rifle – I was the guard who stayed there for all time, watching the prisoners come and go.

Six

MY CHANCE to get to the mountain came with the October boat. Vargas was on board. His impressive face stood out from the others as the skiff came ashore – passive, lean and lined, and I easily imagined it framed by a feathered headdress. He was fifty-eight but as flat-bellied and muscular as I was. The two guards with him, perched on sacks of potatoes and other provisions, were completely the opposite – Roberto, the cook, was an extremely obese, loud-voiced and laughing man, decidedly of Mexican descent. The other, Gerardo, was much younger, slim but with a belly that stuck out like a balloon.

Vargas's black eyes were devoid of expression as we shook hands. Roberto clasped a ham of an arm around me, enveloping me in a stench of sweat, and proudly announced in booming falsetto: 'Meester Gerald, I speek verra bad Eengleesh, no?' Gerardo was the one I took to, and he taught me a lot more Spanish. As he rightly said on one occasion: 'I think you understand Roberto better when he doesn't speak English.'

When Roberto heard that I had been a wine grower, he beseeched me to make him some. I chose stinging-nettle wine as I knew that could be drunk after three days. On the fourth day I served out a half mugful each. Everyone declared it good stuff. We all had another half mugful at supper time.

I was then sleeping on my own in the rear shed, the two couples sleeping in tents. Roberto came creeping in with two mugs. 'Please, Meester Gerald,' he whispered, 'two leetle beets for Vargas and me.' I filled the mugs almost to the brim.

Roberto and Vargas slept in two cabins annexed to the shed and separated from my sleeping quarters by a sheet of corrugated iron. That night, the usual loud snores were accompanied by the most violent of farts.

The wine was finished on the second night. 'More! More!' Roberto pleaded. I decided to experiment with an evil concoction of my own inventiveness – five gallons of rice and lemon wine. I made Roberto wait a full week for it. It was really strong stuff and I rationed it out in tots because it was thick like liqueur. Every night Roberto came in for a 'leetle beet more' for himself and Vargas. One night he returned four extra times with only his own mug. I gave him good measure each time, hoping to knock him senseless.

Everyone was in bed. Suddenly, Roberto's anything but dulcet tones began to serenade the whole camp. 'Shut up', 'Go boil your head' and similar phrases from the guards had no effect. In fact, Roberto's voice grew louder. Abruptly, there was silence. 'Good,' I thought, 'the bugger's out cold.' But I was wrong. Hearing movement, I looked towards the open doorway and there, silhouetted against the moon, was Roberto, mug dangling in one hand, and a fishing line, baited with an enormous chunk of meat, in the other, coils trailing on the ground.

'Come, Meester Gerald,' he shouted, 'wee go catch beeg jack (shark).'

'Piss off, Roberto,' I growled.

He took two or three swinging, staggering steps, caught his feet in the fishing line and pitched on to his face in the sandy floor. He was still lying there, snoring, when I went to sleep. Before morning he had crawled off to bed. We had to get our own coffee and breakfasts. Roberto refused to be woken, and Vargas was as surly as hell.

Like Carlos, I was not at all attracted to Vargas. He loved to be waited on hand and foot too much, and would often lie on his mattress under the almond trees, reading the Bible with one eye and leering at the two girls in their bikinis with the other. While the younger guards' gazes were overt and appreciative, Vargas's were covert. Many times at the dining table he would grimace and complain that his shoulders or neck ached, and

Anne would massage him (she is now a fully qualified masseuse). He got the idea after seeing Anne massaging me after a strenuous paddle against a stiff breeze on the one occasion her fiancé came out with Gerardo and me, fishing. 'That was a real pain and a bore,' he said, yet I was the one with the knotted back muscles. With all due respect to him – and I did feel sorry for him because of it – his body, in shorts, resembled a matchstick with the wood shaved off. The guards promptly named him 'Gilligan' and that was his name all the while he was on the island.

In the third week after Vargas's arrival, when Anne had got him in a good mood with a purposefully long massage, I spoke to him about the mountain. 'It's a bit silly,' I ventured, 'being on this island and never exploring it.' He grunted. 'I've been told,' I went on, 'by previous guards that you would be the best guide and companion to help me get to the mountain.' I could see he was obviously flattered. 'The others come, too?' he asked. 'No,' I said, 'they have no interest in exploring.'

'What about Anna? She is your woman.'

The mistake was nothing unusual. All the guards thought that Anna, as they called her, was my woman. It was because we were often seen talking or walking along the beach alone together. Sometimes, she would sit next to me, with her arm around my shoulders. The guards simply didn't understand companionship between a male and female. There were many innuendos made to me, one being that I shared Anna with Gilligan.

'Vargas,' I said, 'Anna is Gilligan's woman. They are engaged to be married.' I don't think he believed me. 'Have you a map, Mr Gerald?' I fetched it, laid it out on the sand and orientated it. How I wished I had my compass. But that was in the freight. Vargas studied the map then looked at several high points above West Wafer in the direction of the mountain. He studied the map again. With his finger, he traced several routes. 'We go this way,' he said, finally, and ran his finger along West Wafer's small river for the equivalent distance of a hundred and fifty yards, then veered right to go straight up the seven-hundred-foot-high ridge.

'Rain forest plateau on top,' he said. 'Good walking.' His finger went on across the plateau, over small valleys, each higher than the other, along the tops of ridges, through more valleys until reaching the continuously ascending ground of the massif with its two summits; the further one peaked, the nearer flat. They were joined by a lower ridge, sheering down into deep valleys on either side.

'We go right of the first summit,' said Vargas, 'and across the ridge. I have seen that part of the island from a ship. Very bad jungle and very wet.'

'When do we go?'

The usual answer came: *'Mañana.'*

'Yippee!' I cried.

Vargas looked at me, sombrely. 'Mr Gerald,' he said, 'we take no rifle – only machetes. No sleeping in the interior. Too dangerous. Many rock falls and landslides. We cut until four-thirty every afternoon, then return to camp. If the weather keeps fine, we go out every day until we get to the mountain.

That night I was so excited I 'stole' two extra tots of wine to put me to sleep.

Such is the terrain of Cocos that even the large valleys between the high ridges are not flat but undulate with smaller ridges, hills, hillocks and knolls, creating smaller valleys, denes and dells, all interlaced with streams, some of which are mere trickles or bank seepages. With the constant ascents and descents, it is probably quite reasonable to say that any distance walked should have three-quarters of the map distance added; but only if that distance were walked in a straight line, and that was impossible, so always more yardage needed to be added. The map distance between Wafer and the far peak was two and a quarter miles.

The vegetation on Cocos changes with location and altitude. Outside the rain forests, the undergrowth is exceptionally strong and the richness of the many shades of green is quite often dazzling.

In one place I would see a riot of delicate, lace-like fronds and branches; in another, a slim palm, its trunk standing high off the ground on thin, straight roots, spread its plumes above a thick mass of laurel. Tree ferns kept a respectable height under trees, but in open spaces rose to as much as ten feet. The trunks and branches of the giant trees were covered not only by the grass-like epiphytes but by orchids and bromeliads, some of which were protected by garlands or lianas and vines with a maze of tendrils. Thick hibiscus ran along the river banks and fronted parts of the Wafer and Chatham shorelines.

Vargas told me that the tree on stilts was the 'Bow and Arrow' tree. 'Very useful for my ancestors,' he said. 'They made bows from the trunks, and the roots were natural arrows.'

Climbing with one hand seeking grips on rock, earth, branches, ferns and grass, leaving the other free to wield the machete, it took us three hours and forty-five minutes to get to the top of West Wafer's ridge. We crossed two grass-swathed ledges, flushing two turkey-like fowl on one and grazing deer on the other. 'Good hunting place,' said Vargas. In later months, when the trail, like the Chatham-Wafer one, had become well-trodden, it still took three-quarters of an hour to get to the top, and the same time to get back down. The safest ways of descent were either to face the hill or use bottoms as brakes.

After crossing the broad, rain forest plateau we came to the first of the short, deep valleys. The sea and coastline had long since disappeared. Each valley was lush, dense, ant-filled and contained a clear-water stream. I asked Vargas what would be the procedure if we encountered an anaconda or large python. He gave me one of his rare smiles which sent wrinkles running from every line in his face.

'A big snake coils itself in readiness on a low branch and, if there is more than one thing passing underneath, it always drops on the last one. As I am in front, why should I worry?' Then he laughed and brandished his machete. 'Have no fear, I would make short work of any snake with this.'

'Doesn't your arm ache, swinging that thing all the time? Mine does.'

'I can cut for ever!' he cried, and vigorously attacked a clump of extremely tough grass.

Half way through the second valley he looked at his watch as he had done frequently for the past thirty minutes. 'Four thirty,' he said. 'We go back.' We didn't hurry overmuch and were in camp at six fifteen.

Next day I took my map, but it was impossible to say exactly where we were – there was no recognisable, distant feature to be seen. We calculated we were a third of the way, though we had gone off on so many tangents and there were so many valleys that we could be facing in any one of a compass's westerly combinations. The one thing we did have was Vargas's sense of direction.

It was on the following day I had strong cause to doubt it. We suddenly saw the sea and the wide curve of a bay to our right. We gulped down the fresh air, very welcoming after the dust and pollen of what we had been cutting through. The sun hit us like a red-hot iron on our backs. We looked at the map.

'That is this bay,' Vargas declared, stabbing with his finger at the second of two bays on the north side not far from the massif. I disagreed, saying it was the first. 'No,' he insisted, 'we couldn't see the first bay because we were on the other side of this long ridge when we passed it.' I reckoned we were now standing on that ridge, but he was adamant and I relented. After all, he had got us this far.

I was very thankful when we left that oven-temperatured maze of valleys, in which the ants bit unmercifully, and entered a rain forest once more. The terrain began to change into something we had not previously encountered. We were continuously ascending and there was no light at all to be seen through the trunks ahead, which meant that the ground was still rising. Everywhere was wetness and moss. We could see, smell and feel it. Moss-clad trees took on a ghostly, menacing appearance, their branches reaching upwards and towards us like the claws of a witch clad in tattered garments with threadbare sleeves. In places we walked on squelchy moss carpets. Fallen trees, in various stages of decay and shrouded in moss coffins, often blocked our path. Some we climbed over. Others crumbled when we touched them. We were surrounded by a clinging silence as though in a centuries old and forgotten necropolis. The saddest line in poetry came to mind: 'And no birds sang'.

No more were pencil beams of sunlight piercing the top foliage. Large droplets of water fell upon us and clumps of tree ferns soaked us as we sliced off their heads. The thought of being lost and abandoned there filled me with claustrophobic dismay. I was thankful that our exit route was clearly marked. We both shivered, untied our shirts from around our waists and put them on. They were sodden but kept in our body heat.

'We are on the mountain,' Vargas declared, triumphantly. I agreed but was sure we had not passed the first summit. The doubt as to which bay we had seen still nagged me. For another half hour we continued up, the trees becoming closer together and the tree ferns much thicker. Many times we were forced to take a circuitous path. Then we could see daylight ahead and we quickened our pace.

The top was dense with trees and ferns, blocking all view. We sliced through the ferns. As the last one fell, we knew we were on the highest point of the first summit, for there was the main peak about four hundred yards away across a valley that dipped to obscurity.

Vargas was sadly disappointed about his mistake, but I told him he was brilliant to get us to the mountain without a compass.

I had an instant impression of a pyramid, its top truncated on a level with us by a black cloud from which gallons of water streaked in shades of brown. Gusts of chilling wind brought some of the rain to us. The cloud sped on and for a moment we saw a sinister cone packed tight with black-green trees. Then another cloud raced in to release its load. Behind it, from the south, rain-laden clouds queued to land as I'd seen jet liners do when approaching Miami airport from over the Everglades.

I felt rather despondent and not a little dream-dashed. Ducks would be the only happy inhabitants here. Certainly not the sun-loving Incas. Campbell had got it wrong. But no more so than other writers – official, authoritative or otherwise – in their descriptions of and statements about that tiny island's size, shape, height of mountain and distance from mainland. Discrepancies were legion. 'Isla del Discrepancy' would be a good name for it.

We cut down more ferns to give us a view behind of the island's main central valley plain, and a blue sky far away where Chatham's massif lay. We knew the sun would be shining in Wafer.

'Vargas, do you think it ever stops raining here?'

He shrugged. 'Costa Rica's driest months are January and February. Perhaps it might be better then, Mr Gerald, if you want to set foot on the other peak. But there is nothing there.' That was something else he was to be proved wrong about. We were looking at it, but we couldn't see it.

'We are probably the first people ever to stand here,' I said.

'I am sure of it,' he replied. 'Always I have kept a look-out for signs that someone else has been here before – and no one has – at least not for the length of time these present trees have stood. A man would have to make cuts and signs or he would quickly be lost. I have seen none.'

Vargas pointed with his machete to where an arm of the massif ran to the south, forming a large 'vee' with another ridge on the coastline. 'Esperanza,' he said. Bay of Hope. Perhaps that bay held hope for me – not to find the Lost Incas but to establish a home away from the guards. It was my only hope, for I had not been at all enamoured with what I had seen of Chatham, and there was no other accessible bay on the island.

The place where we stood was aptly christened 'Look-out Point' and the trail named 'Kingsland-Vargas'. Many guards and visitors used it. Probably it is still used. The walking time was four and a half to six hours for the return journey, depending on weather and an individual's fitness. Bare-footed, I was able to run its entire length.

Those three days of cutting, returning to camp and going out to cut again really spelled out the eight years' difference in our ages. Vargas slept for three days afterwards, during two of which Roberto took him his meals to his cabin. Naturally, Anne was frequently summoned for massage.

'Oooh!' she said. 'He's like a spoilt, selfish brat.'

'Then don't do it,' I told her.

'It does give me good practice for what I've decided to do when I leave here, I suppose,' she said, resignedly. 'Besides, there's already enough bad feeling in the camp. You know, of course, that everyone is just a little bit wary and frightened of you?'

'Frightened of me?' I said, in astonishment. 'Why?'

'Because, Gerald, you are becoming rather bitter and aggressive, not at all like the easy-going, cuddly bear I first met. This island has changed you, and I don't think for the better.'

We were walking along the beach, far from camp, after she had joined me to collect oysters. 'There's nothing wrong with the island,' I said. 'It's some of the people living on it that get on my tits. You four might just as well not be here for what good you are.'

'I know we must be a terrible disappointment to you and that you call us unadventurous "townies", but you're not doing too badly. You are the one the guards look up to and call Mister, you are Shark Man and you are the one they trust with the rifle. To all intents and purposes, the island is yours – no one else seems to want it. And when your sons are here, you can live wherever you like. Just forget about us.'

'I'll try,' I said, kissing her cheek. 'I'll really try. Anyway, I'm more concerned with getting to Esperanza at the moment.'

I never took anyone to my bowered glade among the Genio, and there I would sit, thinking of a way I could get to the Bay of Hope. Vargas was absolutely emphatic that he wouldn't venture out of camp again. If it came to the pinch I would just disappear one day with the rifle and to hell with their permission.

The plants in my garden were now quite robust. Unfortunately, so were the weeds. Anne helped me with the weeding, often on our hands and knees, and we staked the tomatoes. One morning, Gerardo came to me in the garden with the pleasing news that Carlos was returning on the November boat. Would he be my companion to Esperanza? I wondered. I noticed that Vargas didn't share my joy and he mumbled something about Carlos's being a hot-headed tearaway.

The boat was a week late because of the weather. For those seven days it stormed and blew, white-capping the bay's waves and making the surf so frighteningly high that launching the boat was impossible. I built shelters of palm leaves to protect my plants, but I lost a few and the leaves of the tomatoes turned yellow.

Hundreds of migrating swallows were forced to land on Cocos for shelter. Wet and bedraggled they huddled in large groups everywhere, a favourite place being round the buckets on the floors of the two latrines. So thickly did they pack themselves in each night that the heat-seeking ones below surely died of suffocation. Every morning we threw a dozen or so tiny, stiff bodies into the rubbish pit to be devoured by rats. Sometimes a swallow would succumb in mid-air and crash to earth. Vargas suggested we blew up their tail feathers and throw them into the air. Some would carry on flying. Others thudded back to earth and we would kill them to put them out of their misery.

When the weather broke, the hot tin roofs were covered with black blankets of swallows, sun-bathing, spreading their wings and feathers to absorb the life-giving rays. Like swarms of nectar-hunting bees leaving the hive, they would launch themselves on fast-flying, insect-catching flights and return to the roofs.

Two beautiful, large Cocos hawks, similar to buzzards, were seen perching on a nearby tree to survey the tempting cornucopian banquet spread on the roofs. I sent several bullets by them to frighten them off but they managed to pluck a meal or two. After three days of sunshine, the swallows took off to resume their journey, probably to enjoy the summer in South America. We estimated that they left about a hundred of their fellows behind.

Because of the continual wind and rain, we were thrust much more into each other's company. And more and more I resented the other couple. I knew that soon I would have to do something about their presence, and do it through legal channels and not with a machete.

But one afternoon I found that Anne had a problem, too. Everyone else was either sleeping or reading, and Anne and I sat in the dining room, watching the bay's wet and windy scene. We hadn't spoken for a while and I sensed there was something on her mind.

'Gerald,' she said, eventually. 'I'm a bit worried. I haven't had a period since I've been here.'

'God! It's not me, is it?' was my instant, unsympathetic retort.

'Don't be daft,' she said. 'I had a period the week after you left me in San José.'

'Gilligan?'

She shook her head, emphatically. 'No! No one. I know I'm not pregnant. But I am worried that something's gone wrong inside me.'

'Oh, don't be silly,' I said. 'You're as fit as a fiddle. It's probably due to the water we're drinking here – or tension – different climate – Roberto's cooking!' She laughed at that. 'Come on, cheer up. I just know there's nothing wrong with you.' I gave her shoulder an affectionate squeeze. 'Do you know, I think I must be pregnant, too. I have this terrible craving for a slice of bread.'

'Oh, yes,' she said, 'wouldn't it be heaven to eat some bread again?'

'And so you shall – this very afternoon. I watched Rosemary bake it in Italy.'

'But there's no oven.'

'Ah,' I said, 'I've thought how to get round that. If you want to make yourself useful, get the kitchen fire going.' The fire was a simple construction of empty forty-five gallon drum, both ends removed, draught holes punched through the sides and a grid on top. There was a stack of dry almond wood by the side.

In the store was a two-hundredweight sack of flour that had stood there for five months and only used for soup thickening and batter. It stank of rats and was full of weevils. I broke through the crust and found it quite floury though a trifle damp underneath. Digging down I scooped up several large, double handfuls into a kitchen sieve.

'I hope your hands are clean,' Anne called. 'Don't worry,' I said, 'kneading dough is marvellous for pulling shit out from under the finger-nails!'

With the shaking of the sieve, weevils and a few rat turds came to the surface and I threw them out. 'Ugh!' exclaimed Anne. It took a while to get the flour through the sieve. Eventually, all that was left in the bottom were balls of flour, one or two dazed weevils and hundreds of their yellow eggs. Many of the latter had gone through with the sieved flour. 'Shh!' I whispered. 'Don't say anything. It's all good protein.'

While I searched outside and found three fairly large stones of similar dimensions, Anne put water on for coffee and fermenting the yeast. There was an old, large saucepan in the kitchen and I put the stones in the bottom of it. I put the yeast into a mug, poured in warm water and added a dessertspoon of flour. 'The flour always improves the fermentation,' I explained. I used only a third of the sieved flour to make the dough in a basin and added a pinch of salt. Then I kneaded it, greased a metal plate, put the dough on it in the shape of a bun, covered it with a towel and left it.

While we waited for it to rise – and it did so very quickly in the tropical heat – I told Anne that I had decided to go back to England for Christmas and bring the boys back immediately. 'I think it's only fair they should spend

67

Christmas with their grandmother. She loves them very much and looks after them very well.'

'I have decided to spend Christmas on the island,' she said. 'I've promised myself that. Then I think it will be time for me to go. I definitely shall if nothing has happened inside me by then.'

'What about the other two? And Gilligan?'

'They want to stay here for Christmas as well – but we don't want to spend Christmas with the guards. We are thinking that we might go over to Chatham for the festive days. The rats can't be any worse there than they are here.'

'I never did tell you about the first night here,' I said. 'It was quite funny, really. You know how those bloody things come raiding as soon as it's dark?'

'And how!' she said.

'Carlos, Joni and I were sleeping up there above the food store – the rats' main target. Suddenly, the rats were everywhere, screeching, scratching, running along the rafters within inches of our heads. It was when they ran over us that we couldn't stand it. We got torches and machetes and started swiping at them. Pinto went absolutely berserk below and managed to scrunch a few that leapt from the beams. The other four guards were sleeping in the cabins and they all dashed out with the din we were making and joined in the sport. Chunks of wood flew out of the rafters, dents appeared in the roof, rats' eyes and bare white arses gleamed in the torch beams. Alfredo grabbed a rifle and blew a hole straight through a saucepan and the roof – that's that hole over there. In the end we just gave up. We were too exhausted to do anything. We covered our faces with the sheets and must have looked like three corpses. I awoke with a chewing noise right in my ear.

'I slowly reached for the torch and machete, lying on the floor by my bed, aimed the torch at the noise and snapped it on. And there was a bloody big rat – sitting on its haunches on a rafter next to my pillow, holding a strand of spaghetti in its front paws and nibbling it like a squirrel. Its eyes flashed red and it stopped chewing. Then it wrinkled its nose and took another bite. God! I swung that machete with all my might, missed and dented the corrugated iron with one hell of a clatter. The rat leapt straight on to Carlos,' I laughed as I remembered the scene. 'Carlos sat bolt upright with a scream as if he'd been shot, and his scream made Joni screech as well which brought the other guards out again. I felt a real twit. I didn't know them very well then, you see.'

Hearing Anne's laugh, Vargas came out of his cabin, red-eyed and chewing an obviously revolting taste in his mouth. 'Coffee?' he enquired. It was also a demand. Anne poured him a mug. I told what I was doing. 'No oven,' he said, bluntly. I ignored him. He was content to make sheep's eyes at Anne, in any case.

Without re-kneading I put the inflated mound of dough on its plate in the saucepan, resting on the three stones so that it was insulated from the direct heat underneath. Then I put the lid on with a rock on top to ensure it was airtight. Anne told me when ten minutes were up and we had a look. Vargas peered in, too. The loaf was coming along nicely. I turned it over so the bottom wouldn't burn. Five more minutes and it was done.

Before it had the remotest chance of cooling, Vargas demanded a slice. I buttered it and handed it to him. He blew on it and took a bite, spat it straight out, catching Anne with a few spittle-soaked flecks. 'Disgusting!' he exclaimed. I tasted a piece. 'It's marvellous,' I said. 'Not sweet enough,' Vargas whined. 'I like my bread sweet.'

A few others had arrived for mid-afternoon coffee. They agreed with Anne and me that it was good, and we ate the lot. Vargas insisted I made a sweeter loaf. I did so, putting in a dessertspoon of sugar.

Anne looked at me, sympathetically, when Vargas declared it was still not sweet enough. 'Let the bastard eat cake,' I said in English. Anne also spoke in English: 'Absolutely, make the sod sick.'

She giggled as I mixed in the same amount of sugar and flour. It rose alarmingly like a balloon, sweating and stinking like the proverbial barmaid's apron. 'Put it in the saucepan, quick,' said Anne, 'To kill the yeast.' It turned out heavy, doughy but cooked. 'Yum, yum,' said Vargas, in equivalent Spanish, and he hogged the lot. He never did know what Anne and I were laughing over. All the guards wanted me to make more bread, but I'd had enough so I made them their favourite promise instead – '*Mañana.*'

Seven

PINTO WAS first out of the skiff, gave me a lick of welcome and raced up the beach to the sheds. Carlos and I performed an embracing dance in the sand. 'I couldn't get your freight out of Customs,' he apologised. 'They need your signature.'

'No matter,' I said, and told him about my going to England to fetch my sons. 'I'll bring the freight back with me.'

'Good,' he said, 'I have permission to stay here for several months. You know I want to get married, and we get more money for being here, so I will be able to save.'

I also told him about the trek to the mountain. 'The interior is not quite as bad as it looks, and whether you like it or not, you're coming with me to Esperanza.'

'Anything to keep me away from Vargas,' he said. Then he introduced me to a tall, smiling and large-mandibled Negro in a most remarkable way: 'This is Neg-row. He's a detective.'

'My name is Victor,' said the Negro, with a broader smile. 'And you are Mr Gerald. I have heard a lot about you.'

'Never believe everything you hear,' I returned the smile. As soon as we reached the dining area, Victor took off his tunic and sat at the table with his elbow on it, forearm upwards and his hand open, invitingly. 'Come, Mr Gerald, I hear you are strong. Challenge me.'

I sat opposite him. 'If you insist,' I said as the others gathered round. Victor had probably enjoyed a soft life too much. I beat him first with the right arm and then the left. 'Good,' he said, more pleased than resentful. 'We shall be very good friends.' And we were. Later, I asked him why he was always addressed as Neg-row. 'It is the

71

done thing in Costa Rica,' he told me. 'It is accepted without insult because there in no insult intended. Simply the colour of skin.'

'What do you call me? *Blanco*?'

'Whites are referred to as Gringo, but that is usually for the Americans and is slightly derogatory. No – English are *Ingles*. Chinese are *Chine*.' He pronounced it Cheeny.

'I saw an amusing piece of graffito in San José: "Gringo go home – and please take me with you".' That really tickled him.

Carlos and Vargas never so much as acknowledged each other, and while Carlos liked Anne and Gilligan, it was obvious he didn't take a shine to the other couple. 'Oh dear,' Anne said to me, 'more animosity.'

'Don't worry,' I assured her, 'as soon as we're sure about the weather, Carlos and I are going to Esperanza.'

'If you are going to Esperanza,' Victor interjected, 'you'll be needing this.' And he placed a police-issue compass on the table in front of me. 'And, as it's mine, I shall accompany you.' I was overcome with delight and surprise. 'Victor,' I said, 'you are more than welcome.'

It wasn't until ten days after Pinto received his machete wounds that a recognisable change in the weather occurred. Tension between Carlos and Vargas was electric. One morning, Vargas staggered into breakfast, looking ghastly and saying he was terribly ill. Carlos got up, went to the other shed and returned with a spade. 'What are you going to do?' I asked. Vargas looked at him with weak interest. 'Dig a big hole,' was the astonishing reply. 'With luck, he'll be dead by tonight.' He looked straight at Vargas. 'Die, you bastard,' he said, and went off towards the garden. I wanted to laugh, but decided it wasn't ethical. To my amazement, I found Carlos actually digging a hole! 'Are you serious?' I asked him. He grinned. 'I really would like to dig that bastard's grave, but I'm digging a rubbish pit – the other one's full.' The shock to Vargas was remedial – he made an instant recovery.

On the day the weather made a decisive change, I saw Victor walking alone on the beach and decided to have a private word with him. 'I think we'll be able to make a start tomorrow,' I said. He nodded. 'I think so, too.'

'Victor,' I plunged straight in, 'I know you realise that I have rather a problem with the other couple – should I make an official request to have them removed from the island?'

He stopped and faced me. 'No, Mr Gerald, decidedly no. You are the leader – you received the permission to come here to write a book – therefore they are your responsibility not ours. They have done nothing wrong and should my government be asked to intervene they would probably request you to leave also, for you would be showing that you failed as a leader in your choice of companions. Catch-22, Mr Gerald. It would be better for you, personally, to be friends – if not, then why not simply ignore them? As it is, they keep very much to themselves; either way, it shouldn't be too difficult for you.'

'Thank you, Victor, I must be very dense not to have seen that aspect of the situation myself. But I can't be friends with them – so I shall ignore them.' And that was precisely what I did.

'A word of warning, Mr Gerald,' Victor continued, 'during my two weeks here I have detected many things – that is what I am paid for – and I know that Vargas is very jealous of you. He doesn't like Carlos but you do, and Carlos respects you. But Vargas holds more sway in the department than Carlos does.'

'And all I wanted was to be a second Robinson Crusoe,' I sighed. 'Ah, well, such is life as the saying goes. Anyway, thank you again. Incidentally, were you sent here to investigate relationships?'

He regarded me coyly. 'That would be telling, Mr Gerald, that would be telling. But I promise you, you do have my sympathy. Come, we'd better make ready for tomorrow. I am looking forward to our journey very much.'

Carlos had already begun the preparations. The two dining tables were covered with provisions: rice, black beans, tins of meat and tuna, hard biscuits, butter, condensed milk, coffee, sugar, packets of vegetable soup, cooking oil, pepper, salt, plates, mugs, cutlery, matches, candles, toilet paper, frying pan and saucepan.

Vargas came in, stopped in amazement and asked, sarcastically: 'How long are you going for – a year?'

'It does seem a lot,' I said, easily, before Carlos could flare up, 'but three men can put a lot of food away in a week.'

'You don't need half that much,' said Vargas, airily. Carlos began packing it all into a large rucksack. When he'd finished, we could hardly lift it. Vargus sat there with a supercilious 'I told you so' look.

'I think we had better take two rucksacks,' I suggested. 'No one man is going to carry this lot, and there are still blankets and groundsheets to consider.' Victor fetched his large rucksack and we divided the weight equally between the two, putting three blankets on top of the food in one and tying three groundsheets in a roll on the bottom of the other. There was still quite a bit of weight in each.

When we discussed our route, Vargas again intervened. Carlos immediately walked away. I listened to what Vargas had to say. He advocated the Kingsland-Vargas trail, then making our way down into the valley and following the river to the bay. 'The central plain is all thick jungle – a lot of hard work,' he said. 'It will be very hot. No rain forest for cool, easier walking.' It made sense to me. I measured the two routes on the map. The direct way from bay to bay, transversing the island across the plain was two and a quarter miles, compared to three and a quarter miles if we went via the mountain, but we would then only have to establish a trail for one and a quarter miles from the look-out. Contour lines across the mountain's valley and river were almost touching in places, indicating high waterfalls, but balanced against the hot, dense and ant-filled jungle of the plain, plus the impossibility of a straight route, I heeded his advice.

There was another decision I was called on to make: should we take Pinto? I said yes – his wounds had healed to thin, dry scabs and he would probably be miserable and pine if Carlos left him behind.

Ebullient with the adventure awaiting us, we set off at eight o'clock in the morning. 'Let's blaze a trail across this island once and for all,' I said.

'To Esperanza!' Victor yelled, as though we were bent on seeking El Dorado.

'Good luck,' Anne called to me, waving. 'Take care.'

Carlos let out a gleeful war whoop, and Pinto barked wildly at being allowed out of camp once more.

It was one of those rare days when the sun shines on the mountain, causing steam to rise everywhere, in some places resembling plumes of smoke from a hidden camp-fire, producing in me a fleeting hope for the Lost Incas. As we made a brief exploration to determine the best way down I thought I saw a flash of white mid-way up the cone. 'Fairy tern,' I told myself, as there were many of those beautiful, snow-white and dove-like birds on the island, which frequently hovered just above our heads. Many times had I reached up a hand to stroke them, but always they just evaded my touch.

I elected to be the first one to cut while the others carried the rucksacks. As I sliced through the foliage I felt extreme elation with the hot sun on my back. Like Vargas, I declared I could cut forever. Often, Carlos and Victor would sit, waiting for me to clear a good distance ahead, and frequently call out to ensure I was all right and not tired. Suddenly, I was seized with an egotistical desire to cut all the trail singlehanded, and I did so despite their efforts to take over.

That part of the island was harsher and more rugged. Landslides were evidently more frequent, recent and heavier. Three times I came to yawning holes left by the roots of huge trees torn from the ground, their trunks ripping out boulders and flattening other trees as they had sped down until checked by the piles of debris before them.

I jumped down the last three feet into lush grass by the side of the pure and sparkling river, running shallow over multi-coloured stones and pebbles and swirling by sentinel rocks. We laid full-length in the water with Pinto, drinking. He'd kept to my heels for most of the four-hour descent.

'I don't much fancy having to climb back up there,' I said. They agreed; nonetheless, we cut two poles and stuck them in the ground to show where the trail started. The river changed after a hundred yards or so and we passed through a series of rock gorges, their pools opaque and green and each one deeper than the last until we were forced to swim, floating the rucksacks before us on simply-constructed rafts of poles tied with vines. The last pool was quite long and produced in us a feeling of unease at not knowing what lay beneath its surface. Carlos exclaimed once that something big and slimy had touched his leg, causing us to swim faster. Pinto troubled us by trying to climb on to our backs and inflicted weals on our shoulders.

The river broadened and shallowed, then entered a short canyon which tapered to a dark rock tunnel through which the deep water ran with force. There was a narrow ledge above the water level on the right-hand side and we proceeded along it with Carlos in the lead. I brought up the rear, Pinto at my heels. The dark walls streamed with water and the surface of the ledge was very slippery. The noise of the rushing water was amplified to the decibels of thunder.

I was carrying the rucksack which contained the blankets. Stopping momentarily to ease the weight on the shoulder-straps I felt the dog hit my calf. He slipped and fell into the racing water. I shouted and half-turned to grab. The rucksack hit the wall and pushed me into the torrent as well. The weight of the rucksack took me straight to the bottom, then the current lifted me up again. I broke surface to hear Carlos's and Victor's screeches and I had one of those resigned, calm feeling that I had often experienced in Korea that 'this was it' – Sod's Law would decree that a five-hundred-foot waterfall waited at the end of the tunnel. Instead, first the dog then I were spewed from the tunnel and swept halfway across an enormous, shallow bowl. I got to my feet in knee-high water and dragged Pinto to the side. My hands shook as I examined his wounds. All the scabs had been washed off in the other pools. Thankfully, the flesh was still white and there was no bleeding.

'I bet you think we are really silly buggers to put you through all this,' I told him, and received a rough tongue between the eyes. Carlos called out to me from the tunnel's exit. 'Are you all right?'

'Just a bit shaken,' I shouted back. They waded over and we sat at the water's edge. The sun was very welcome. 'We tried to catch Pinto,' said Carlos, 'but he was out of our reach and we couldn't see you at all. We thought you were both dead.'

Partly in shock, I called Vargas every name bar his right one, much to Carlos's amusement. 'One thing's for sure,' I said, 'we'll go back across the plain. By the way, where's the compass, Victor?'

Victor took it from his trousers pocket and opened the lid. It was full of water which he tipped out, then laid it open in the sun.

'Listen!' said Carlos, suddenly. From downstream we caught the sound. The faint rumble of cascading water. 'I'll go and investigate while you wring out the blankets,' said Carlos.

Twenty minutes later when he hadn't returned we began to worry. 'I think I had better go and see what's happened,' I said, with awful thoughts filling my mind. As I exited from the bowl I saw with relief Carlos wading back round the next bend.

'Where have you been?' I demanded.

'I had to have a good shit,' he said. 'It was giving me a headache.'

'Jesus Christ!' I said, despairingly. 'Anyway, what's ahead?'

'No good,' he said. 'Very deep waterfall. Cliff on both sides. No way to get down – and there's no sign of the bay. In fact, there's no sign of anything except treetops.'

'Where do you think we are, Mr Gerald?' Victor asked.

'Lost,' replied Carlos, and spat.

'Don't be silly,' I said. 'How the devil can we be lost

when we're following the river?' I took the map from its plastic case. 'Look – here's the first waterfall so we're about halfway along the river.'

'Is that all?' asked Victor. 'I thought we were almost there.' His waterproof wristwatch showed five minutes to five. There was no chance of our reaching Esperanza before dark at six thirty.

'We'd better think about making camp,' I advised.

They agreed. 'But not here,' said Carlos. 'Too dangerous. We must get to high ground because of falling rocks.'

It was dark before we reached sufficiently level ground to lie on, with stars shining down through the trees. Groping our way about, we cut ferns and made beds. Too hot and tired to think about eating, we spread out our groundsheets and damp blankets over the ferns and laid on top of them.

'One thing we can be thankful for,' I said, 'it isn't raining.'

'Shush!' they both hissed, as though I had tempted fate. Sure enough, within ten minutes it was pissing down. 'Now look what you have done, saying that,' Carlos groaned.

We turned our bedding over and got between groundsheet and blanket, pulling the latter high so that the rain didn't belt into our ears. I could still see the humour of it when I fell asleep.

'Mr Gerald! Mr Gerald!' Carlos's insistent whisper brought me awake. 'Eh? What?' The rain had stopped. Carlos and Victor were sitting up. A small, intensified spot of light beamed an inch or so from my nose. I sat up. We were all bathed in a deathly, grey–white, the perimeter a circular wall of unknown blackness. 'What are they?' asked Victor, nervously.

'They're glow worms,' I said. 'Haven't you seen them before?'

'No,' they both replied. The glow worms had clustered round us and it was their cumulative light that made it appear as though someone had spotlighted us.

The night's stillness was suddenly, startlingly broken by a boulder or some large object hurtling down, crashing through branches and undergrowth to our right. There was silence, then the echoing thud of its striking a new resting place far below.

'Very dangerous,' breathed Carlos, in a small boy's voice.

'We're OK,' I said, 'there's no high ground immediately above us.'

We sat, straining our eyes into the darkness beyond the glow worms' light. I thought how it wouldn't be too difficult to convince oneself, as Campbell had done, that unseen eyes were out there. If not the Lost Incas then perhaps a strange, ferocious animal. But not according to Pinto. He was cuddled up in Carlos's bed, intent only on sleep.

I had difficulty in waking Carlos and Victor in the morning. They had preferred to sit, talking for quite a long time, listening to my snores and more rock falls. 'This landscape must be changing all the time,' said Victor.

'It's bound to,' I replied, 'with all the volume of rain.' But we couldn't find any water where we were, so it was pointless building a fire for coffee. We ate a cold breakfast and moistened our mouths with rain drops left in the corrugations of fern leaves.

I opened the lid of the compass. There were already streaks of rust and the needle didn't swing until I tapped it a few times. South was where the only light could be seen through the tree trunks. We went over to have a look, and we were completely overawed. The view alone was worth my going to Cocos.

We were on the spur of a rock precipice overlooking an enormous concave of rugged beauty. On the far side, hundreds of coconut palms bent their heads in stark relief against the Pacific's pre-sun-up grey.

To our far right, soundless through distance, a band of white water fell a thousand feet, striping the black cliff face of the mountain's southern arm. Mid-way down, the cascade struck a tree-fronted ledge, disintegrating into spume and spray, shooting back up behind the trees and flying high over their lofty tops. It then re-formed into a wider band, falling the second five hundred feet to disappear into a tree-clad ravine.

Behind us, cloud poured from the mountain's cone like a stream of black smoke from a chimney.

To our left, in colourful contrast, the bright red, white and brown of the cliff on which we stood, curved in a giant arc to merge and disappear in a criss-cross of verdant valleys. From the valleys, a long, forested slope ran gently down the coast to the palms.

I cannot recall the many inane comments of wonderment we exchanged.

'It's so wild and magnificent,' said Carlos. 'And it all belongs to us – Costa Rica.'

'It is easy to imagine why many pioneers who encountered scenes like this in their travels, decided to stay and abandon their homelands,' I said. For a moment, I even wished to be Costa Rican, for I envied their proudness of ownership.

We were gazing at God's own country which, in my insignificance before its untamed splendour, I felt, could and would annihilate any man who had no mechanical assistance. The rough-hewn floor below appeared even more harsh and cruel than anywhere else I had seen on the island, with its obelisks of rock continuing up from hill-tops, their needles protruding high and sharp through the surrounding foliage.

I pointed to the far, coastal slope. 'I'm afraid that will be our only way down to the bay,' I said, 'and I think it will take us nearly all day to get there.'

'We are seeing a sight that no one else has seen,' said Victor. 'I thank you, Mr Gerald, for bringing me here. I shall always remember this view. But as you say, we still have a fair way to go.'

We reached the coastal slope shortly after two o'clock but it was impossible to see the sea because of the high vegetation. We could, however, pin-point our position because we recognised the strata of the cliff face on which we had been standing, and the mountain was also in view. Although the sun beat down on us, rain was falling on the massif.

We emptied both rucksacks, spreading our provisions

on the two groundsheets and hanging out the blankets. Our choice of meal was tinned tuna on buttered biscuits, and water from a nearby stream. While we were sitting, munching, I cast my eye over the food store.

'Carlos,' I said, 'how many tins of condensed milk did you pack?'

Victor instantly looked uneasy. 'Three,' said Carlos. 'Why?'

'There are only two.' Carlos made a closer examination. 'But I know . . .' he began.

Carlos looked at me, then we both looked at Victor. 'Bastard!' Carlos screamed, and drew his machete. Victor scrambled to his feet in alarm.

'Put that bloody thing away, Carlos,' I said. 'You know as well as I do that you'll never use it. Sit down and shut up, I'll handle this.' I was angry, too. I also felt a little humiliated by the fact that a man I had sought advice from should turn out to be a thief. And a police detective at that. But whatever his position in Costa Rica, out there in the jungle there was no class or distinction.

'Why?' I asked him, simply.

'Because I was starving,' he replied, sullenly.

'But a whole tin?' I said, incredulously.

'I couldn't put it back in the rucksack. It would have run over everything.'

'When did you eat it?'

'At that very steep part where we nearly broke our necks. You and Carlos went on ahead to find a way down.'

I felt disgusted with him and allowed my voice to show it. 'Next time you're starving, let us know and we'll all eat. We're in this together. I know you're not as fit and used to this sort of thing as we are, but Christ, Victor, that was a pretty low thing to do. Supposing one of us had an accident – that tin of condensed milk could be vital. As it is, the way things are going, we're going to need every scrap of that food – shared equally. Equally. Understand?'

'I'm sorry,' said Victor. Which was really all he could say.

As the afternoon wore on we found ourselves once more competing with sun-down. 'I don't want another night out here,' said Carlos.

'We'll make it,' I said. 'It just can't be much farther, now.'

'Mr Gerald,' Carlos groaned, 'you said that an hour ago.'

I was right this time, for suddenly, joyfully, we saw the Pacific not far below us as we made our way down what must surely be the steepest river in the world, our descent being by leaping from boulder to boulder. But we quickly saw there was something wrong. All the palms were on our right, not immediately in front of us. In absolute dismay we found out the reason when we emerged at the river's narrow, cliff-sided mouth – the river had its own private exit, separated from the actual bay by a massive, insurmountable rock against which the unchecked force of the Pacific smashed with earth-shaking violence.

Our sighting of the sea had brought new life to tired limbs, and raised our spirits. Now we sagged with disappointment and knew that from somewhere we had to command and conjure up more energy. Less than thirty yards were between us and the Esperanza's shore, but there was no alternative but to go back up the river to the nearest ridge, cross over it and come down again. We could see that this tiny beach was completely submerged at high tide.

With every back and leg muscle aching, we dragged our bodies and the rucksacks up the river's boulders for a hundred yards, then pulled ourselves to the top of the ridge with every obliging branch.

Long since, only Carlos and I had carried the rucksacks. Victor could hardly carry himself. We half-stumbled and half-ran down the other side, and blissfully joined Esperanza's very surprised freshwater shrimps, small fishes and other forms of life in the green, algae-bottomed river.

We lay there, with the cool water rushing over our shoulders and looked up at the palms. I counted as many as a hundred and twenty-one trunks before giving up.

There was every size of palm from sprouting nut to lofty, laden tree. Beneath them all, interwoven and connecting, was a thick bed of palm leaves and fallen nuts.

'So this is how coconuts grow when man doesn't interfere,' I said.

'Yes,' said Carlos. 'No one has been here for many years, and no one has ever lived here.'

'I think it's real spooky,' said Victor. 'Not a nice atmosphere at all.'

'That's probably because you're tired,' I said.

'It's because he's a Neg-row,' said Carlos. 'Those buggers still believe in witchcraft. What do you think about the bay, Mr Gerald?'

'I don't think I'd want to live here. Bit too closed-in for my liking; but I'll wait until tomorrow to see what I really think.'

Esperanza is shaped like a long-stemmed, narrow-mouthed coaching horn, its front to the sea and its long tail leaning against the mountain. That tail now disappeared into settled cloud which blotted out the summits and was creeping down to us. Carlos and Victor made a tent-like shelter of palm leaves. Driftwood covered the beach, and I built a fire for cooking. While the soup was heating I cooked rice, Costa Rica style: fried lightly in a smear of oil, then adding water to swell and simmer the rice. Evaporation left each grain separate.

Torrential rain, strong wind and darkness came together. The rice was instantly water-logged and the soup diluted. We sat in the shelter, miserable, cold and wet, drinking our first hot meal, which wasn't, and listening to the deluge on the roof, the whistle of the wind and the roar of the surf.

Victor crouched nearest the entrance. In silhouette he looked like a cannibal as he voraciously attacked a previously opened coconut with his large jaws. Carlos nudged me. 'First, he eats the coconut,' he said. 'Then us.' We both thought it a huge joke, even if Victor didn't.

Our blankets had been soaked again when the rucksack

had dipped many times into the 'steep' river. Water came through the roof and flowed under us. Pinto gave a long groan and stretched out in the only dry spot. 'Bugger this for a game of soldiers,' I said in English. Then, in Spanish: 'Can't you do something about the water? I'm going to light a fire in here.'

'OK,' said Carlos. 'Come on, Victor!'

An almost full moon showed through a misty brown blanket of cloud under which heavier, black clouds scudded towards the mountain. Carrying a groundsheet I made my way through the palms to the beach, the driving rain like a breath-catching cold shower. In alternate dark and half-dark I made a heap of wood on the flapping groundsheet, gathered up the four corners and staggered back. Carlos was piling more palms on to the shelter while Victor dug a trench with the machete behind it.

I started a fire with two broken candles, the smoke dissipating through the leaves. Pinto never budged. In the small flame's light I could see that no water was coming in, and I yelled the good news to the others.

'Now for a nice cup of coffee,' I said. Carlos stuck the saucepan outside, wiped it round with grass, rinsed it and quarter-filled it – all with rain water, so hard was the downpour. I complimented them on their building capabilities, they praised my fire-making. Wisely, they said nothing about my cooking.

The fire and coffee soon made us warm and drowsy, and we slept soundly till morning. The fire was still alight and the rain was still coming down. It poured all day. The only time we went out was to get more wood and relieve ourselves. Pinto accompanied us only once, and he refused to eat a thing. 'There's something very strange about that dog,' I remarked.

'There's something strange about this whole bay,' said Victor. 'Even in daylight it's menacing and sinister.'

I was beginning to agree with him. Carlos didn't like the bay, either. 'When the rain stops,' he began. Victor interrupted: '*If* the rain stops . . .' 'All right,' said Carlos, broodily, 'If the rain stops we'll go back to Wafer. Until it does, we'll never be able to climb out of here. OK, Mr Gerald?'

'OK,' I said. All of us hated the thought of the journey back. We even discussed the possibility of building a raft and floating back round to Wafer. We dismissed that idea when, with groundsheets beating against our heads and shoulders, we viewed the surf and the barriers of jagged black rocks. 'I don't think there are many days when anything could be launched or landed here,' I said. 'One thing's for sure – I'm not bringing my sons to live in Esperanza.' I wondered what poor, lost and anguished soul had named it Bay of Hope.

For two more days and three nights we were penned in by the rain. It was during the second night that I awoke as though something had disturbed me. A misty moon shone through a temporary break in wind and rain. There was only the sound of the surf. A man suddenly appeared beneath the palms, and I felt the hairs on the back of my neck prickle. I sat up as he began searching for something. He was dressed in baggy trousers, tucked inside calf boots. Round his forehead was a red sweatband and he carried a short sword. I realised then that I was dreaming. No I wasn't, or I wouldn't be conscious of a hard piece of rock sticking into my left buttock. I moved to the side and did something I had only read about in books – I pinched myself three times in different places. Seeking more proof to show that I was awake I touched an ember, and instantly stuck my burnt finger in my mouth. The man was still there, striding purposefully under the palms and poking with his sword. I dismissed the thought of waking the others. They were nervous enough as it was. 'This is ridiculous,' I said, laid down and closed my eyes. But I just had to look again. The man was closer. He turned and looked directly at me.

His eyes were brown, his face kindly and round, and his nose straight. I felt no fear at all, yet had no inclination to greet or go out to him. For how long I watched it is difficult to say. The man turned about abruptly, marched

85

down to the beach and disappeared through the palms. He never returned and I never saw him again.

One of the first things I noticed in the morning was my burnt finger. I never mentioned it to Carlos and Victor but often reasoned with myself why the man's face had looked so very much like the face of the first man I shot in Korea.

On the fourth morning when I awoke, Carlos had already made coffee. He was smiling. 'No rain,' he said, happily. I went out. On the grey Pacific's horizon, long black strata of cloud were visible in the dawn light, the only sure indication I know of fine weather. We quickly ate a usual breakfast of boiled rice and sugar, and used the last of the condensed milk in our coffee. Because we'd had nothing much else to do except eat, there was very little food left. Yet we had all lost considerable weight. Pinto raised his head and weakly flapped his tail, but still he wouldn't eat anything.

Victor dug a deep hole with the machete and buried the rubbish while Carlos and I started packing. We decided to leave all utensils behind. Then a miraculous thing happened. As soon as Pinto realised we were leaving he transformed instantly from sick, dying dog to skittish pup. We just couldn't believe it. 'Carlos,' I said, 'there is something very wrong here in this bay. That dog never even went outside, and never relieved himself.' As though to prove my words, Pinto stood with hind leg raised, pissing up a palm, for an incredible time, interrupting the flow for a leviathan bowel movement.

The compass was now very rusty, but the needle still swung and constantly showed north without dithering. 'We march on twenty-five degrees,' I said. 'We miss that steep river completely.'

At six twenty precisely, according to Victor's watch, we set off with nary a backward glance. Before we had gone twenty yards, Pinto had disappeared over the ridge. We cut marks on trees but sliced only the minimum of undergrowth, in some places only sufficient to squeeze through. It felt luxurious to be pouring with sweat again. The plain was everything but, the biggest obstacles being the maze of hills. We crossed over the smaller ones but went round the high ones, making the necessary bearing adjustments.

Stopping only once for a meal of the last two tins of meat, we virtually force-marched across the hot, ant-biting plain. At 5 p.m. we reached the source of the Genio's west fork. Downhill all the way, we walked into camp at six fifteen. 'Impenetrable jungle' and 'dense interior' notwithstanding, we had crossed the two-and-a-quarter-mile, narrowest width of Cocos in eleven hours and fifty-five minutes.

I had a right to feel proud – I had linked two bays – and I had done it with a rusty compass that I had become extremely suspect of when we were half-way across but, fortunately, didn't pack up completely until we saw the recognisable feature of an outstanding clump of black–green trees on the top of a hill above Wafer.

Eight

TELEVISION NEWSREELS were showing people being mown down by machine-guns on the frontier when I arrived in San José with my sons and an English girl acting as their nanny. I was advised to register their names at our embassy in case the situation became such that we had to be taken out in a hurry. I was informed that His Excellency Michael Brown, British Ambassador to Costa Rica, wanted to have a word with me.

'Ah, Kingsland – good to see you,' he said, with all the sincerity of a diplomat. 'I'll get straight to the point. This chap, Blashford-Snell – heard of him?'

I nodded. 'Yes, though perhaps he's not quite my type.'

'Yes, well, probably not. Anyway, he's making Drake's voyage round the world all over again. I understand that his ship, *Eye of the Wind*, has been refused entry to the Galapagos. He's now thinking of calling in at *Isla del Coco*.' He allowed himself that short piece of fluent Spanish. 'If he does get his permission, perhaps you'd be good enough to act as host – take him and his crew along some of your trails and that sort of thing.'

'Of course,' I said.

'Splendid. Good to see that the British spirit of adventure is still alive. By the way, it's damn queer about that aeroplane on Cocos, isn't it?'

'Aeroplane?' I exclaimed. 'What aeroplane?'

'Haven't you heard? They've discovered a crashed plane on the island – on your mountain, to be precise.' I forced back an expletive. He was going on: 'Bit of a mystery – it's believed to be an American bomber posted missing in 1942. Good job you cut that trail. I understand that US Air Force investigators are going over with you.'

I was impatient to get to Puntarenas, but he ordered

88

coffee and we discussed the island and the local political situation. As he showed me to the door, he said: 'Sorry you didn't find the Lost Incas. I think Campbell was rather a romantic, you know. Get your boys to have a good look for the treasure – it's certainly a fine adventure you're giving them. Good luck.'

The first person I bumped into at Maritime Police Headquarters was Vargas. He's been on the pre-Christmas boat to the mainland with me.

'Who found the plane?' I asked, instantly.

'Carlos and another guard,' he replied, with a trace of bitterness.

'Do you know exactly where it is?'

'Right opposite Look-out Point,' he continued, sadly. 'The other guard had binoculars.' I knew how he felt.

He told me that two MTBs were going to Cocos when the Americans arrived in a few days' time. The boys and I were booked on FP407 with him and a change of guard. FP405, a three-engined hundred-and-twenty-footer, was taking the US officers, two park rangers and a hundred and fifty bags of cement. The more astonishing news – and rather unsettling – was that a permanent radio station and a post office were to be established in Wafer, and the park rangers were to be a permanent fixture.

As the politicians say: a wind of change was blowing. It is so often the case when one goes away for a few weeks. I had been away for just over three. Thank goodness I was going to live in Chatham.

A senior officer beckoned Vargas as I enquired about Anne. 'Gone,' Vargas called, as he made for the office. 'They left on the boat which rushed to the island when the plane was found.'

Colonel Guillerme Martí, chief of staff, placed a police jeep at my disposal, and I got my freight from Customs and put it on the quay. I also took the jeep to get four months' provisions, lace-up boots for the boys and a new machete for me. For Christmas, the boys had received Swiss Army knives and fishing tackle.

Colonel Martí told me that the plane was believed to be a Flying Fortress.

The weather had changed dramatically in the short time I had been away. Puntarenas was even more like an oven and the sun extremely vicious. Heedless of my warnings, the boys had dashed down the crowded beach to get into the comparative cool of the sea, and within twenty minutes their shoulders were blistered. I had to buy special cream from a doctor. Soft drinks also cost me a small fortune. Rory wrote to his grandmother: 'I don't like the place because it is too hot. The place is falling to bits and it smells of rotten smells that linger in the air night and day. The only good place is in a bar, where it is very cool, and you drink Fanta and Coke all day.'

The old – and, I thought, long dead – resentment towards US fighting personnel returned when I saw the US Air force officers on the quay when we were called to board. They were dressed to kill – literally. Full combat suits and holstered pistols. 'No nuclear warheads?' I asked. Memories of their fellows, strafing our lines in Korea, killing eight of us, brought the sarcasm. But in the hustle, bustle and general chaos, I had no time for conversation. I had three boys to count, constantly, and to ensure that they and our freight were not left behind. 'I shall be glad when I get you on that island,' I said. 'At least I'll know where you are.'

But the captain was more remiss than I was in looking after his flock. We were almost out of the long, wide harbour when he realised that a member of the crew was missing. It was Gerardo, who, with Joni, was now employed on board. We had to go all the way back to fetch him. By the time we returned to our original position, FP405 was a dot in the distance.

'Plopsy, are they always like this?' Roddick asked.

'I cannot tell a lie,' I said. 'Yes.'

A smiling Joni and Gerardo joined us as we sat on the deck with our backs against the bridge. 'I was just telling my sons what a bunch of cowboys you lot are,' I said.

'It's much too hot for discipline,' Gerardo replied.

Naturally, we chatted about their days on Cocos and the sharks. Gerardo still had his balloon belly, which he patted fondly. 'Do you remember that day you took me to Chatham, Mr Gerald?'

'I certainly do,' I said, looking at his stomach. 'You hung on to almost every sapling up the hill, coughing your lungs up. Still smoke as much?' He nodded. 'Afraid so.' Still remembering, he went on: 'And what about the pig you shot? I never saw so many flies.'

In English, I told the boys how, on the way to Chatham, I had killed a small sow, eviscerated and decapitated it, and hung it in a tree for our return. When we returned, two hours later, we couldn't see it for flies, so we left them to carry on eating. Three days later, when I went back there on my own, hunting, the pig had gone.

'Joni,' I said, 'that very first afternoon on Cocos will always stand clear in my mind. You just walked up a palm and brought down the green coconuts. Pipas, you call them, and we sat on the beach, belching like mad after we had drunk the juice of several. I wanted to tell you a story about a band of pirates on Cocos but I couldn't speak Spanish then. The pirates drunk so much pipa juice that they became too drunk to stand and had to be carried back on board!'

They laughed. 'And what about that first deer we saw which ran into the sea?' said Joni. 'We saw that big fin. Then the deer disappeared. I felt a little bit frightened then.'

'So did I, Joni. So did I.'

Three-quarters way out of harbour we repaired aft out of the way of the bow wave we knew would sweep the deck. I have never failed to be thrilled when a fast boat suddenly opens the throttles after crawling at harbour-regulation speed. The bow rises high with the powerful surge and the spume flies astern. I saw by the boys' faces that they felt it, too. 'This is the life, Dad,' said Rory.

Within two hours, although the Pacific was as flat as a pancake, they began to look a bit peaky. 'Get your heads down,' I told them, and Joni and Gerardo kindly gave them their bunks.

Somewhere, somehow in the night, FP405 went adrift. She was captained by the one most notorious for getting lost! We arrived at Cocos at seven thirty. FP405 arrived one hour later. The island looked even more tropical, sparkling and appealing. It felt really good to be back and absolutely blissful to walk bare-footed again.

Rubbing liberal quantities of oil on the boys, I told them to keep in the shade as much as possible. Their morning was complete when Carlos and Joni took them off hunting. I went to inspect my vegetable garden. It was an absolute shambles. Produce had been taken without care, no one had bothered with watering during the month it hadn't rained, and the plants looked as if they had been trodden on. All that work, I thought.

The guards were older and more officious, and the two park rangers appeared watchful and hostile. I sensed antagonism between them and the police. The rangers told me they had orders to exterminate the pigs, rats and wild cats. 'Killing off the pigs sounds a bit silly,' I said. 'They're good, cheap food.'

'*Señor*,' was the clipped reply, 'this island is a national park. Wild pigs are not indigenous. They were put here by pirates.'

'What about the deer? They aren't indigenous, either.'

'I expect we shall exterminate those, too, in time.'

'In that case,' I said, 'I'd better eat as many as I can.' Their reply was to walk away from me.

Although I really wanted to accompany the party to identify the aircraft – as I was asked to do – I knew I couldn't. It wouldn't be right to subject the boys to that journey in unaccustomed heat, and I had no wish to leave them in Wafer. Besides, I wanted to get ourselves settled in Chatham. Both skiffs and both crews took all day unloading the cement, and it was late afternoon when the boys and I were dropped off at Chatham. They had slept for two hours after their 'fantastic' hunting trip, on which Joni had dropped a deer.

A disappointment to me was that the deep-sea diver and his girlfriend were still on the island and had taken a tent with them to live in Chatham. I had misinterpreted Vargas when he'd told me 'they left on the boat'. Only Anne and Gilligan had departed.

Within a few days, the boys and I were in complete estrangement from the couple, though we all made an effort to be friends. To make matters worse, the nanny, who had told me just a few days before we flew to London that she was only sixteen – causing me to get a hurried letter of consent from her mother – was intensely disliked by my sons. 'We are not taking orders from her, Dad,' they said. 'She's only as old as we are.'

After about a week in Chatham, she went to live with the couple. From then on, the two camps and occupants kept their independence and distance – though we occasionally shared a good catch of fish.

It had been Rosemary's idea that I took a nanny, and a friend of hers wrote an advertising blurb for me in the *Evening News*, bringing about forty replies. Several were very suitable but had to give lengthy notices to employers. When I made my final choice, I thought she was a young woman of at least twenty-one!

While I hurriedly cooked supper, the boys made parcels of the most vulnerable of our provisions and hung them on thin cords from the rafters in the fishermen's hut.

Before going to England I had spent a week in Chatham, having a look round and tidying up the hut. Victor joined me for the last couple of days for 'peace and quiet'. He told me that really bad quarrels had taken place between Carlos and Vargas after I left. During that week I had hammered a few flapping wall planks back on, nailed pieces of wood over the rat holes in the floorboards, and strengthened the four wooden bunks. Then I had cleared the hut's surrounds, freeing a young lime bush from strangling vines, and finding several empty forty-five gallon drums. One of them I used for a stove after discovering some rusty grids. The hut's roof was constructed of corrugated bituminous sheets, and there were stacks more under the raised flooring.

But, as Victor agreed, the row of closely grown, thick-trunked almonds in front of the hut and their huge, lichen-clad boughs, gave the already black-painted hut an even blacker, oppressive character and atmosphere. A small river ran close by, and I singled out a flat piece of land, covered with thick hibiscus, on the opposite bank, right on the open shore. It was there that I decided the boys and I would live.

The bay's shoreline was divided by a rock promontory, the end of which was always submerged, and I found it was an excellent place for fishing. In readiness for when the boys and I arrived, Victor had helped gather and cut a store of firewood.

While at first I had felt excitedly emancipated to be away from the guards at long last, I was soon experiencing moment of loneliness and melancholy. It was then I real-ised there were two kinds of loneliness – mental and physi-cal. Although I'd had people around me I had been lonely within myself ever since Rosemary left. Not even the nearness of my sons had alleviated it. One didn't need to go to an uninhabited island to experience that kind of loneliness.

Foolishly, I suppose, I had reasoned that loneliness was loneliness, and, as I was already lonely, I wouldn't miss the company of others. But I did; and I rebuked myself for it. 'You're a bloody fine Robinson Crusoe,' I said. That I *was* physically lonely evidenced itself on the night before Victor arrived. A white, three-masted ship in full sail suddenly appeared as I sat on the promontory, scaling and gutting an Australian big-eye. The sun was dipping fast.

She was heading in and I hurried back to put coffee on for my guests. I watched the ship through the doorway and, to my dismay, she suddenly turned her stern towards me. She was going out to sea again! I rushed to the almond trees, the surf was thundering on their roots, and began waving and shouting. Small chance I had of their seeing me there in the impending dusk, let alone hear me. From the hut, I kept my eyes on her: her position such that she was visible between two almonds. A top-mast light was switched on and long after the ship itself had been swallowed by the descending darkness, I saw her mast light. Sometimes, hopefulness caused me to imagine that it was bigger and brighter because she was coming back. Then the light vanished forever.

I felt too miserable to eat more than a mouthful or two of fish, and that night I was engulfed in extreme desolation as, once more, I pulled my sheet over my head and went to sleep with the rats.

Rats were, of course, the main topic of conversation that first night there with my sons.

'Why didn't we bring the air rifles, Dad?' Rory demanded. 'I can't get to sleep with all their noises.'

'Plopsy, what would they eat if we weren't here?' asked Roddick.

'Birds' eggs and baby birds, I expect.'

'Well, why don't they sod off now and eat some?'

'Dad?' Redmond's voice came to me from below. 'Can I sleep with you?'

'Poofter! Poofter!' sang Rory.

'Yes, if you want to,' I said. 'Then I want all of you to shut up and go to sleep.' But they didn't. Not for a long while. Neither did I. But not because of the rats. I was thinking of the parting words of their grandmother: 'It's a very big responsibility you are taking on, you know. If anything happens to one of those boys, it will be a cross you will have to carry for the rest of your life.'

Nine

OVER AND OVER again I tried to convince myself that I had no real cause to fret unduly about the boys. 'Children are survivors,' I said; and I firmly believed that William Golding's island depiction of them in *Lord of the Flies* was a pretty accurate one. I'd taught my three to shoot in Wales, ride horses in Italy, swim in Switzerland, and they had often fished alone off the coast of Sardinia. I had allowed them to roam the wild Tuscan mountains and to walk the lonely moors and forests of the Berwyn Mountains in North Wales with their individual guns, and they brought back rabbits, pigeons and game. So why should I worry about them just because they were on a partially uninhabited tropical island?

'You are not, repeat not, to go into the sea on any account,' I found myself ordering them on that first morning. 'Not even for a paddle. Understand? If you want to cool off, use the waterfall behind the hut – and watch out for the freshwater shrimps – they have enormous pincers which really hurt.'

Every so often I stopped whatever I was doing to see where they were. 'This is ridiculous,' I told myself. 'I'll be a nervous wreck in a week.'

'You're like an old mother hen,' Rory complained. 'We know what we're doing. There aren't many boys of our age who can do the things we can do.' One of the first things they did was make spears of saplings and eight-inch nails and go up river. Within half an hour they were back. A triumphant Rory had a disgusting-looking black fish on the end of his spear. About fifteen inches long, it had no scales and its body was covered in mucus which made it almost impossible to hold. It also had long whiskers. Our fish book told us it was a catfish and delicious eating. It was, too, and it fed all four of us.

The next thing they brought back was an old lobster trap, the wooden frame rotten but the wire cage serviceable. I told them there was a Costa Rican legend that the Cocos Island lobster would never enter a trap, but they insisted we tried. About a hundred yards out, in front of the hut, was a huge rock shaped like a barn and, after re-newing the trap's frame, we placed it just in front of Barn Rock at low tide just before dark. We jammed it between two submerged rocks and put the catfish's head inside for bait. Next morning, when the tide was out again they rushed to see what they had caught. No lobster but a baby white-tip shark! I showed them how to kill it and fried it for lunch.

On the second attempt, they proved the legend wrong. It was a good sized lobster, too. I 'drowned' it in the river, then boiled it. As we tucked into it with a pinch of salt, I couldn't help but remark that a wine and brandy sauce was decidedly needed.

On the following two mornings, after baiting the trap by torchlight because of the successive, fifty-five-minute retardation of the tides, we caught, firstly, a fat-bodied moray eel, and, secondly, a plump sea bass. While the taste of the former was to our liking, the needle-sharp bones were not. Reddick aired all our feelings when he said: 'I think we'll give eel a miss, Plopsy.'

On the fourth morning – disaster. The trap was chewed to pieces. In one part of the twisted wire of the cage was the remains of what had been a fair-sized Australian big-eye. 'Jaws?' Roddick asked. ' 'Fraid so,' I said. 'Wow!'

Later that day, Carlos arrived, bursting with news about the aeroplane. 'It *is* a Flying Fortress – a B-22 – and definitely the one reported missing in 1942. It has the name Fury painted on the side in big letters.'

From the evidence of the wreckage, the plane had belly-flopped, ripping a pathway through the trees and breaking into many pieces, some of which would take a mammoth clearing operation to find. 'But most of the plane is in the one place,' said Carlos, 'and the other pieces are too small to worry about.'

'What about the crew?'

'You know what this island is like, Mr Gerald. Of the eight men, we found only a shin bone and a few foot bones in a flying boot.'

'I wonder what a Flying Fortress was doing in this part of the globe in 1942?' I wondered. 'That was the third year of World War II.'

'They didn't say,' replied Carlos. 'Here, these are for the boys.' He handed me three live rounds of machine-gun ammunition. 'The belts were still in the guns, but the barrels were bent.'

Looking round, he said: 'I see you have been very busy.' We had already cleared the ground and pitched the two tents. Over coffee, he told me that the Americans had gone back to Puntarenas, and the trails had been much easier without the rain. 'Even the horrible one down the valley wasn't too bad,' he added.

'What was it like, going up the peak?'

'Terrible – and very steep. The plane hit the mountain sideways, not full on, and the path it made was completely covered. The jungle there is much worse than on the other summit.'

'Good job I didn't take the boys,' I said, but felt a pang at not being there.

'By the way, the trail between here and Wafer is very good and very clear now. You and the boys are more than welcome to come over.'

'Carlos, thank you,' I said, 'but first we've got to work, building our hut.'

And work the boys did for those first three weeks, swinging picks and axes, digging out rocks and roots, cutting poles high up in the rain forest and sliding them down, digging holes for the uprights and putting on the roof timbers. Most afternoons, the boys slept for an hour or so, and even I dozed. Still there had been no rain and, if anything, it was getting hotter. In turn they would go to the promontory to fish. Rory, the keenest fisherman, claimed the biggest catches – a shark longer than he was, and a conger eel which sunk its teeth into the heel of his boot before he flattened its head with a rock.

The two park rangers paid us a visit, obviously to see what we were doing. Inspecting our seed boxes, they asked: 'Where did you buy the seeds?'

'Puntarenas,' I said. I knew they couldn't object to that, nor to the trees we had cut for our house and the raft I intended to make. I had previously secured permission to do so.

'After this, no more trees are to be cut – and no more trails to be made. We have told the police so as well.' Refusing coffee, they picked up their rifles and left.

'Plopsy, I don't think they like you.'

'They are just doing their duty, Roddick, that's all. I can't say I envy them. Come on, let's get on with it.'

When the roof was half-finished, Vargas came over and asked if I'd look at the paraffin refrigerator FP405 had left for them. Only Roddick and Redmond wanted to go. Rory complained he'd had too much sun. I made him promise he would not, under any circumstances, leave the camp. Before we reached the top of Green Hill I had fearful doubts about leaving him behind. It took me a while to decarbonise and trim the fridge's wick and adjust the flame so that it burned blue not yellow. I refused their invitation to eat, and we hurried back. We'd been gone nearly three hours. My worrying had been in vain. A very proud Rory was waiting to greet us. Singlehanded, he had finished the roof. 'You little bugger,' I said, in relief and pleasure. 'So that's why you wanted to stay behind.'

Next day, the dry spell broke with a violent thunder-storm. Water trickled through Rory's roofing. I saw why. He'd put the nails through the corrugated sheetings' troughs. His brothers teased him. 'Never mind, Rory,' I consoled him, 'it's a bloody good roof, it only leaks when it rains.'

When the storm had blown itself out next morning we patched up the roof. 'Let's make it a really beautiful home, Dad,' they said.

'OK. How about planting coconuts and bananas as well?' There was only one coconut palm in Chatham, which we christened Toulouse-Lautrec because of its stunted appearance, so we trooped over the ridge to Wafer and brought back four plants and two banana cuttings. We planted the bananas at the rear of the garden, and the coconuts at each corner of the house.

With no help from me, the boys built three hammock-like bunks with poles and corrugated sheets, one side secured to the central uprights and the other suspended by twin ropes from the beams. I thought they were ingenious. There wasn't room for a fourth bunk so I stayed in the tent. Strangely enough, the rats left the boys alone, the only intruder being a scorpion which Rory killed the instant he felt it on his thigh.

With the completion of the raft, we were, in fact complete – house, garden and transport. 'There is one thing missing, Dad,' said Redmond. 'What's that?' I wanted to know.

'Mum,' he said. I think they understood why I suddenly walked away along the beach. I could see Roddick having strong words with Redmond.

The launching of the raft was one of our proudest days. Looking at the surf, Roddick put on his British Army officer's voice: 'Come along, chaps, this is a dangerous mission, and some of you won't be coming back. If you should succumb, dear old Plopsy, I shall bury you in front of Toulouse-Lautrec.'

'Stop pissing about and push,' said Rory.

'He said the P word, Dad,' said Redmond.

'Shut up, the lot of you and push,' I commanded.

Four men could have carried the raft, but not one man and three boys. We'd waited until the tide was right in and were trying to slide it down the bank by the side of the house. The shore shelved steeply, causing the waves to break twice. They were first shattered only a few yards out by the backlash of the preceding wave. If we could get the raft through that first breaking point, we had only the unbroken waves to contend with.

When the raft suddenly began to slide, its weight and momentum pushed the nose deep into the water, but it quickly rose to the surface and was lifted high by a wave which thudded the rear end into the bank. For security I had linked the raft to a tree with fifty metres of new nylon rope. I held the rope tight as the raft flew up and down.

'Take your paddles and get into position,' I said. 'Roddick and Rory at the front corners, Redmond at the back on the right. Lie flat with your paddles beneath you and hold on like grim death.'

'I say, Plopsy,' said Roddick, 'do you think we're doing a wise thing?'

'Just get on board, Roddick, please. It's bloody hard holding on to this rope.'

'Hurry up, Dad,' they pleaded as they laid there with the waves rushing over them. I wedged my paddle between two cross-members and went down the bank until I was chest-high in the broiling sea but safe between the jutting rear ends of the balsa floats which stopped the raft from crushing me.

'Hold tight!' I called, before the next wave smashed against my face. I gripped the rear cross-member, braced my feet against the bank and thrust hard in unison with the ebb. The raft shot forward, over and through where the waves met high, taking me with it. I swam strongly with my legs. 'Get up and paddle,' I yelled. They scrambled to sitting positions and dug in their paddles. Clear of the bad point, I hauled myself aboard. We'd made it! 'Yippee!' they chorused. 'We're afloat!' I began paddling too.

So pleased were we with the handling of the raft that we forgot about the rope. It snapped taut and shot Roddick and Rory over the front into the water. I pulled them out immediately.

'Plopsy,' said Roddick, 'why don't you ever get everything right?'

'The bloody thing floats doesn't it? I can't remember everything.'

'I'm going to write and tell Grandma,' Redmond said,

teasingly, 'that you forgot about the rope, and Roddick and Rory were nearly eaten by sharks.'

'Do you want me to exterminate him, Dad?' asked Rory.

'Come on, left-hand down a bit,' I said. 'We'll go back in to just before the breakers.' The raft turned with little effort and glided back towards the shore. Near where the tops of the waves began to curl, I told them: 'Right-hand down, lads. Gently now.' And we brought the raft round to meet the sea again.

'Decision time,' I said. 'Shall we cast off?'

'Yes!'

I untied the rope. Roddick made the sign of the cross. 'For what we are about to receive . . .' he intoned.

'*Quo vadis?*' I asked.

'London,' said Redmond.

'Piss off, Redmond,' said Rory.

Around us and above us was clear azure. Behind us, verdancy. We went out beyond Barn Rock, stopping now and then to study the colourful bottom, fifty feet below. We saw several shoals. 'We should have brought the fishing tackle, Dad,' said Rory.

'This is just the test run,' I replied. 'We'll have plenty of fishing trips.'

Carlos and Pinto accompanied us on one. 'Mr Gerald, this is a very good raft,' he complimented us. 'You have three *bueno* sons.' With Carlos there, I ventured further out to near the eastern point and we caught a large sea bass which Carlos cooked and helped us eat.

Ten

THROUGHOUT THE halcyon days of February and March, several yachts came to Chatham. At first, the boys rushed down the beach to greet the shore-boat, proud to show the visitors our house, garden and raft. Then I noticed that territorial possession was taking hold of them as it had of me. Suddenly, the boys had had enough of the ephemeral interruptions to our set island way of life. No longer did they help to beach the visitors' boat in the surf, and we found ourselves showing distant hostility to the people wading ashore.

Relationships in the other bay were deteriorating. 'I wish I could live here with you and the boys,' Carlos said to me on one or two occasions. 'Park rangers and police do not mix.'

The boys, singly or in pairs, had taken to visiting Wafer, often staying the night to be fresher for the return. One morning Roddick came back and told me that a machete duel had taken place and Vargas had made him sleep on his cabin floor out of harm's way. Two nights previously, someone had pressed a knife against Vargas's throat, but he'd beaten off his assailant.

'It wasn't Carlos, was it?' I asked.

'No, Plopsy, he keeps well out of the way.'

'From now on,' I told all three boys, 'you will go to Wafer only when I am with you. And that's a definite order.'

'We don't mind, Dad,' said Rory, and the other two agreed. 'We're happy to stay in Chatham. We love it here. And we never, ever want to leave this island.'

'How many times do I have to tell you? We can't stay here forever – my permission is only for one year – June.'

'But you could easily get that extended,' said Roddick.

'Perhaps I could. But I don't think I really want to. We'll have people breathing down our necks. No – you must go back to school and I shall try to find another island. So make the most of it.'

They certainly did, and lived every day to the full. 'We're never bored here, Dad,' they would often say. We explored all of Chatham's massif, and sometimes we pretended to be treasure seekers, trying to follow the directions on some of my old 'X marks the spot' maps. Unfortunately, many of the starting off places had long disappeared with landslides. Another handicap was that for years, the island maps were printed upside down, so if we stepped out fifty paces north, as one map directed us to do, we would have fallen off a cliff into a hundred feet of water! Redmond came rushing back one day, saying he had found gold way up the river. We went back with him, and he showed us glistening flecks of gold-like particles. I searched around, but could find nothing more.

Treasure seekers' holes and tunnels infested the bays. One, midway along Penitentiary Point, was almost fifteen feet deep. Carlos and I had almost fallen down it, the first time we ventured along there. The boys loved the small caves in the eastern corner of Wafer and we crawled through the small tunnels the treasure seekers had made, linking them. Behind the fishermen's hut, we found a hole with a boulder in the middle of it, with an almost rotted wooden pole still in position where someone had once tried to use it as a lever. Many times I would look at the deep holes in that awful rock and root-filled ground and imagine the despair of the digger, throwing up the last spadeful and finding – nothing.

In the evenings, the boys and I played word games of our own invention: going through the alphabet, naming animal species, countries, rivers, mountains, counties of Britain and the States of America. They were of good educational value, especially a game I knew called 'Botticelli' where one of us had to be a famous person, and the others could only find out who he was by asking questions about other famous persons whose surnames started with the same initial. If Botticelli didn't know the famous person in the question, he had to answer a direct question, stating whether the person he'd thought of was dead, male, a statesman, a German, and so forth.

One late afternoon, when a thunderstorm was approaching across the sea to the north, I looked along the beach to the promontory where three small figures sat, intent on fishing and oblivious to the nearing flashes of lightning. 'Proper little islanders,' I thought. When I reached them, Redmond proudly held up two Australian big-eye. 'Come on,' I said, softly, 'it will be pissing down, soon. I've got the fire going. Let's get those cooked for supper.' It was then I realised just how happy and contented they were on the island, and I felt actual resentment against the government for making it a national park.

During the first week of April, Carlos arrived with the news that six English friends of mine would be arriving on the next boat. 'Six?' I said. 'I only know about two.'

During the Christmas period I had extended an invitation to an architect friend, Richard Evans, and an art student friend of his called John Froy. I was quite surprised when they came ashore. Richard had brought with him an American, Ron Bennett, whom he'd met in Afghanistan, Ron's fiancée, Anne Tippett, and her cousin, Heather, both of whom were Australian nurses. John had brought his girlfriend, Dee Hyde, an art teacher.

I felt rather imposed upon at first, but when I learned they had brought their own provisions and that Ron and the two girls were leaving on the next boat, I relaxed a bit. Ground was cleared and three tents were pitched next to the fishermen's hut.

To my relief, the boys took to all our unexpected guests without exception – and harmony reigned. It was heaven, having someone else do the cooking; Anne and Heather often helped me to invent meals, some of which turned out absolutely foul! But they were really good cooks. I told Heather that I wished I'd known her before. 'It wouldn't have done any good,' she said, 'I couldn't stick this life for very long.'

'Story of my life,' I said, 'all the women I know refused point-blank to come here with me. Pity, really, because I don't think you can ever get the right sort of people by advertising.'

I really enjoyed the evening discussions, mainly round the new dining table Richard had designed and built, though many nights we played word games. Heather helped me to make wine and some good sing-songs ensued.

Almost every day, Dee roamed abroad with her canvas and easel, Redmond, the artistic member of the family, sometimes going with her. Ron spent most of the time chopping firewood to keep himself in trim; and Richard and John loved going out with us on the raft, and soon got used to seeing sharks swimming around and beneath us. 'You've got three brave boys,' said Richard. 'I know,' I replied, 'but don't tell them that.' Like the boys, they were more than a little startled on meeting 'Charlie' for the first time.

After a fortnight's acclimatisation, Richard and John said they would like to see the mountain and the aeroplane. None of the others wanted to go, so we set off with the boys to stay overnight in Wafer. There I recognised trouble with a capital T.

There were now eight guards and four park rangers on the island – it was hinted that the latter would be taking up residence in Chatham. Also worrying was the news in the Costa Rican national papers about public outcry over English people being allowed on the island when Costa Ricans were banned. One headline said: 'Uninhabited Cocos – Come Join the Crowd'.

Besides declaring that six more people had joined Kingsland and his sons, one newspaper stated there were now the following people living on the island: the crew of a scientific research boat, a group of ham radio enthusiasts, an 'unnamed' marine expedition, a Spanish journalist, a North American photographer and a North American science writer.

'Well, where the bloody hell are they?' I asked Vargas. 'They have been here, Mr Gerald. Many boats have come into Wafer that you haven't seen in Chatham. It is all very bad, Mr Gerald. The six people you have brought here will make it bad for you.'

Next morning, we went to the mountain. It was disastrous. No sooner had we reached the base of the peak than the heavens opened. We were climbing a dry, narrow gully which suddenly became a water-chute with rocks hurtling down as well. I was leading, quite a way ahead. Seeing the stones and rocks bouncing towards me, and fearing for my sons' lives, I screamed: 'Get out of the gully! Get out of the gully! Cling on to anything, but get out of the gully!' I looked down and, thankfully, saw them, clinging clear on ferns and saplings. Water gushing everywhere, we made our way back down to where the investigating party had built a simple shelter of sloping, palm leafed roof on six poles. I built a fire and we slept there, cuddled together for warmth. The rain was coming down even harder in the morning. 'We're never going to get up there while this lasts,' I said, renewing the fire. 'Let's go back home to Chatham, Dad,' said Redmond. 'I don't like it here.'

No one did. Why, I asked myself, did this part of the island, when rain fell, give everyone a feeling of despondency and desolation with little incentive or will to do much at all? We all agreed with Redmond and left. On the way back, John almost took his left eye out with a branch. It seemed to affect the vision of his right eye, too, and we had to guide him for most of the rest of the way. It took several days before his eyes were normal again.

A week later, Richard said he would still like to see the aeroplane.

'Count me out, Richard,' I said. 'I feel absolutely drained.'

'Come on, old mate,' he said, 'don't let that lot in Wafer get you down. Against you, they're a bunch of arseholes.'

'It's not only that,' I said. 'I think I've seen enough of that part of Cocos. I have been there three times, you know.'

None of the grown-ups wanted to go with Richard. 'I

will, Plopsy,' said Roddick. I looked at him. He had really matured in the past four months. 'I want to do something for you, Plopsy,' he said. 'What's that?' I asked.

'Stand on that fucking peak and yell.'

He did, too. And Richard took photographs of him to prove it. I wrote a slushy comment in my diary on their return: 'Today, Roddick showed himself to be a man.'

I told him to make notes and then to write an essay about it. One line in particular stood out: 'The view I saw was meant only for people who love loneliness.' He called his essay: 'Roddick's Journal of his Miserable Travel to *Cerro Yglesias*'. That was the name given to the massif. It means hill of churches.

They had taken a tent, which they left there, and a small gas stove, which they broke. Richard had lost his contact lenses. 'I signed my name on part of the aeroplane's wing, then, leaving Richard sitting there I chopped a path for twenty minutes through ferns and razor-grass to the top.'

First fog, then rain descended. 'We went back to the camp and into our tent with our cooking gear and food, and we stayed put for the rest of the day and night. I have now started to feel lonely and sad without my dear old Plopsy and want to get home as soon as possible from this wet, cold and lonely valley.'

Later: 'We cannot get any wood to burn, and as the stove is broken, I made some cold coffee which looked the most disgusting muck I have ever seen. So I gave it to Richard and had powdered milk and sugar. Richard has put on his glasses which make him look very different, and he has read out his diary to me about his stay in Iran which was very thrilling.'

For the first two of the four days Richard and Roddick were away, Redmond had been busily making what he called a one-man raft.

Heather suddenly came running up to me, exclaiming: 'Have you seen where your youngest is?' She pointed out to the bay.

'Oh, my Christ!' I said.

Redmond and raft were no more than a blob, going up

and down on the choppy water and heading for where the hammerheads skulked. Everyone ran down the beach with me. We cupped our hands round our mouths and yelled 'Redmond' at the tops of our voices. He heard us and waved. We jumped up and down, waving our arms and shouting for him to come back in.

I stood on the shore, waiting twenty minutes before he came flying through the surf. At first I was determined to tan his arse till it bled blue beetroot. Instead, I swept him up in my arms, kissed him and told him I was unutterably proud of him. 'But if you ever do that again . . .!' I threatened him.

Ron, Anne and Heather had already begun preparing for their departure when black diesel smoke belched on the horizon and we could see that the gun-boat was heading for us at full speed. She had a cannon bolted on her bow deck and when she turned broadside to anchor, I saw her marking – FP408. She was one of the main marauders of the fleet, almost constantly on patrol. Her crew was a trigger-happy bunch. She had called in once before to Wafer in the autumn and the crew had stood on deck, blasting at every bird they could see. I wasn't at all surprised when nothing fell.

Now, she lay quietly at anchor, making no move to put down her skiff.

About thirty minutes later, FP405 rounded the point from the direction of Wafer. Her skiff came ashore, bearing a dark-haired man in a pin-stripe suit. 'I am looking for the man they call Mr Gerald,' he said. I introduced myself. 'I have some bad news for you. My government has decided that you and your party must leave the island.'

He was not in a position to say why. But I knew, anyway. 'When?' I asked.

'Tomorrow, after mid-day. Some of you will go on 408, the rest will go with me on 405. And, please, Mr Gerald, we want as little trouble as possible.' I held my tongue as he handed me his card. It said: Willie Azofeifa, Official Mayor, Ministry of Public Security.

Early next morning, Carlos arrived, looking slightly

embarrassed by it all. 'I think Vargas put the boot in,' he said.

'Maybe so,' I replied. 'Anyway, I'm donating all the tools, including the Danarm chainsaw to the island – will you look after that?' He nodded and we shook hands rather formally. 'Goodbye, Mr Gerald,' he said. 'I shall always remember you.'

At mid-day, 408's crew came ashore and loitered around, watching us pack the last of our things, and pretending they didn't have rifles.

'I'm certainly not going with that murderous-looking bunch,' Heather declared, emphatically.

'Don't worry about them,' I said. 'You're as safe as houses so long as you're the target.' We finally agreed that Richard, the three boys and I would go on 408.

With the exception of Richard, the others were muttering words like 'insulting', 'degrading' and 'being forced to leave at rifle-point'. They were going to write to MPs and embassies.

'Stop bellyaching,' I said. 'No one's pointing anything at you. And you're getting a free ride with meals to the mainland.'

The boys stood on the stern to get the last glimpse of Cocos. I vowed I wouldn't look, but I did. She was as my first sighting of her: a cottage loaf in a sea of mist.

Rory and Redmond held my hand. Roddick stood behind and placed his hand on my shoulder: 'Never mind, Plopsy. You cut trails and showed them there was nothing to fear on that island – and you found a lost bomber for the Yanks. We'll go back one day and see how our coconuts are doing.'

In San José, I went to see Alvaro Ugalde. 'I am sorry, Gerald, there was nothing I could do. I hope you have enough material for your book.'

'I think I may have material for only part of a book,' I said. 'But don't worry, I'll send you the promised autographed copy.'

Part Two

34 S, 79 W

Juan Fernandez Archipelago, South Pacific

One

IT IS, perhaps, strange how an absurd, irrelevant thought
can occur, cutting completely across a person's set intent
and concentration. I was ascending the broad, stone steps
in front of Armada Headquarters in the port of Valparaiso,
Chile, a country ruled by military government under President Pinochet. At the top of the steps, two alert sentries
in immaculate naval uniform held sub-machine guns at the
ready. There was nothing cowboyish about their dress and
stance. Incredulously, I saw that telescopic sights were
affixed. Not at all sporting! It was then the incongruous
thought flashed through my mind: 'I bet they wait until
some miserable wretch has run a full block, then, just
when he thinks he's safe, they make a sieve out of his
back!'

When I was within a few paces of the portal they
guarded, the two sentries stepped smartly sideways
towards each other, blocking my path. Slowly, I took my
passport from my inside pocket, proffered it and said in
Costa Rican Spanish, which didn't quite match Chilean
inflexions: 'I have an appointment to see Vice-Admiral
Raúl Lopez.' The sentries neither spoke nor budged. A
white-spatted petty officer materialised from between the
pillars and took my passport from under the arch of the
two snub barrels.

'Wait here,' he said. The two sentries resumed their
original positions and the three of us pretended I wasn't
there.

Five minutes later, an affable, English-speaking lieutenant, Ivan Ramirez, led me along wide, polished corridors
to a reception office. 'I understand you have a personal
letter of introduction from Señor Alvaro Pineda, our
cultural attaché in London?' I handed it to him and he

113

disappeared through another door. Another five minutes and a long corridor later, he halted me in front of double doors, gave a couple of hard raps and threw the doors wide. '*Señor* Kingsland,' he announced in a loud voice. He had to shout – it was a barn of a room with a patterned carpet running its full length to a mammoth desk which dwarfed the silver-haired man sitting behind it.

As he rose and extended a gold-ringed cuff his blue eyes twinkled. I warmed to him immediately. His voice and dignity were an English gentleman's.

'What do you think of our beautiful country?' he asked, indicating I should be seated in a splendidly upholstered chair.

'Beautiful,' I replied, and meant it.

We chatted, drank coffee and then I told him what I wanted. He placed both gold-ringed cuffs on the desk in front of him like a shield. 'What you are asking,' he said, 'is for me to order one of my ships to take you to Alexander Selkirk Island?'

'Yes,' I said.

He looked at me, steadily. Then: 'Do you know how much it costs to put, say, a destroyer to sea?'

I told him it must be quite a lot. 'Can't you pretend it's on manoeuvres,' I suggested, 'or let me have something smaller?'

He accepted the joke with a short laugh. 'Leave it with me,' he said, 'I'll see what I can do. Tell Ramirez where you're staying.'

Three days later, Ivan the Terrible, as I openly called him, rang to say that the vice-admiral had pulled a few strings. 'You and your Girl Friday are booked on our best-loved ship, leaving in six days' time. She'll drop you off at Robinson Crusoe Island – that's the best we can do.'

I couldn't believe my eyes – or my good fortune – when I saw the *Esmeralda*. She was a four-masted, Spanish brigantine and the pride of Chile's Armada. A natural for John Masefield's 'Sea Fever', she was certainly a tall ship, for her masts soared a hundred and sixty feet to the stars.

'What more fitting vessel could I have wished for?' I mused as tugs pulled her away from the quay. Quarter-sails were hoisted, then, amid cheers, goodbyes and ships' hooters, *Esmeralda* pointed the nude breasts of her voluptuous figurehead towards the open sea. In a final salute of farewell, a jet fighter streaked low across her bows, dipping its wings, then roared high into the blue.

As in Masefield's poem we were headed for the 'gull's way and the whale's way where the wind's like a whetted knife', for between us and the Juan Fernandez Archipelago was the cold rage of the mighty Humboldt Current which sweeps Antarctic waters northwards.

Like many an exquisite, aristocratic lady, *Esmeralda* demanded and commanded the attention of men – no less than two hundred and twenty of them to care for her three hundred and seventy-two feet of beautiful body-line and twenty thousand square feet of sail-cloth when fully dressed. On this voyage she was also showing a hundred and twenty sea cadets her desires and needs at sea. They were laughing, eager, energetic young men, hauling on the ropes, climbing the dizzy rigging to the topsails and scrubbing the decks. Then *Esmeralda* entered the Humboldt.

Though the sun blazed down, the wind grew stronger, almost violent. Medium to heavy seas drew up and *Esmeralda* took on a permanent list to starboard. Her bow and stern swung crazily, lifting clear of the water as though she were a giant sea-saw, balanced amidships on an imaginary fulcrum.

I watched as the cadets, in ones, twos and threes, staggered to the rails and washrooms. One was thrown from the rigging and taken to the sick-bay with concussion.

Remembering the adage that it is better to spew up something solid, I repaired below to the officers' dining room. Only three officers were there and, after short greetings, we sat in silence as the portholes dropped sharply below the surface, darkening the room, then shot up to allow daylight in once more.

For starters, the waiter brought us *'panedas'*, Chile's

national dish – Cornish-type pasties, containing a mixture of egg, cheese, spices and what appeared to be apple. Now and then, my plate slid away on the white linen tablecloth and I was half-slithered from the leather chair. My three companions were also experiencing the same discomfort and we managed brave smiles at each other.

'Where is your Girl Friday?' one enquired, in fluent English, as plates of rice, vegetables and pieces of roast chicken were placed in front of us. He continued before I could reply: 'She is a very fine English rose. You are a very lucky man to be going to live alone with her on Selkirk Island.'

I grabbed at my plate, missed, and the waiter returned it from the end of the table. 'I have no idea where she is,' I said. 'She was allocated the last berth in the women's quarters and I have to sleep in the sick-bay.'

'Probably the best place,' said the officer. We all smiled.

The waiter filled our glasses with an excellent white wine. I took several hasty sips. The glass was instantly re-filled and my chicken leg began to scream *mal de mer* at me. 'I have been through a cyclone in the Indian Ocean,' I said, 'but it wasn't as bad as this.' In obliging endorsement, *Esmeralda* gave a tremendous shudder as though she had run into a brick wall and she dipped sideways to a more crazy angle. We clutched our plates and the table as all cutlery, wine glasses and condiments went flying. *Esmeralda* returned to her normal slant, pitch and roll and the waiter restored order with a sort of measured, knees-bent, springy step that would have left Nureyev howling with jealousy.

'The Humboldt always produces a different movement in ships,' said another of the officers. 'A kind of gyration.'

'I know,' I said, taking up my new knife and fork in resolute, aggressive mien. 'My stomach is following it.'

'Come! Eat and drink!' encouraged the officer. 'It is the only way.'

Gradually, my ear fluids and stomach juices began to behave. Three more officers joined us. 'I hear you are a personal friend of Vice-Admiral Lopez?' said one of the newcomers. There was a trace of German ancestry in the voice, and a subtle 'Ve haf ways of making you talk' in the probe.

'Not exactly personal friend,' I said, casually, 'but we do get on rather well. Why do you ask?'

'How else could you be an honoured guest on this illustrious ship?'

I let it go at that and suddenly the conversation turned to a subject I was dreading – politics. It was almost like the Spanish Inquisition, but I had plenty of time to weigh my answers, assure them that my book would not be concerned with the seamy side of life, and I almost tugged my forelock whenever Pinochet's name was mentioned. Well, life is very short, and death is for ever.

Luckily, the topic was changed by their wanting to know the difference and connection between Robinson Crusoe and Selkirk, and the re-naming of two of the archipelago's three islands.

I told them that one had to go back to the year 1704 when Spain and Britain were battling for piratical supremacy of the South Seas. Two British galleys, the *Cinque Ports* and the *St George*, under the command of William Dampier, who had great knowledge of the area, sailed up the Chilean coast where immense quantities of Spanish gold were rife for the plundering.

A Scotsman, Alexander Selkirk, was First Mate of the *Cinque Ports* under a Captain Thomas Stradling, and the two men were constantly and bitterly quarrelling about the ship's sea-worthiness and conditions on board.

After unsuccessfully engaging several Spanish ships, Stradling was forced to sail, for re-fitting, to the archipelago's largest island, Mas a Tierra, three hundred and sixty-five miles due west of Valparaiso. There, the hatred between the two men became such that Selkirk demanded to be left ashore with his few belongings. 'And he lived, in complete solitude, on that island for four years and four months,' I said.

In February 1709 he was taken off by a privateer, the *Duke*, captained by Woodes Rogers who said that Selkirk appeared on the beach as a 'wild man, wearing skins of the goats he had killed'. Selkirk had difficulty in speaking because he could remember only half words.

The account of his time alone on the island caused great interest in London, and his story was first published in 1713 by Richard Steele in *The Englishman* magazine. Later, Woodes Rogers published an article about his rescue of Selkirk.

Also in 1713, a journalist called Daniel Defoe was released from Newgate jail after serving a sentence for publishing 'libellous' articles on politics and religion. Selkirk's story came to his notice and, in 1717, Defoe was studying detailed maps of Juan Fernandez. He was also perusing Sir Walter Raleigh's account of a desert island near the mouth of the Orinoco, off the coast of Venezuela.

'What isn't known for sure,' I told my audience, 'is whether Defoe actually met Selkirk, though it is thought that they did because they both visited Bristol where they had a mutual friend.'

In 1719, Defoe's immortal story of Robinson Crusoe was published and it took the world by storm. Although the setting was an island off the north-east coast of South America, the story kept fairly close to the article in *The Englishman*. Defoe never denied that, but the most he ever admitted was that Crusoe's adventures were based on a 'man living today'.

'There was an actual Robinson Crusoe,' I went on. 'He was a fisherman and he is buried in a church in King's Lynn. Defoe used his name.'

Two years later, while Defoe bathed in the wealth and fame his book had earned, and he was busy with his second novel, *Moll Flanders*, the body of Lieutenant Alexander Selkirk was slipped unceremoniously over the side of HMS *Weymouth*. He was just one of many to die of fever off the hot, pestilent coast of north-west Africa.

In 1966, in an effort to attract publicity for Chile and to encourage tourists to inhabited Mas a Tierra, the government changed the island's name to Robinson Crusoe. The archipelago's second largest island, uninhabited Mas

Afuera, lying ninety-six miles further out, was re-named Alexander Selkirk, though Selkirk had never set foot on its shores.

'So you see,' I said, 'Mas a Tierra should really have been called Selkirk Island, because that was where he lived. But, of course, the fictional Crusoe has the world-wide, romantic-escapist appeal, whereas few people have heard of the factual Crusoe, Selkirk. Even I am guilty of not giving Selkirk deserved acknowledgment because I am trying to be a second Crusoe.'

'Perhaps your book will set the record straight,' said an officer.

'I sincerely hope so,' I replied. 'That is the main reason why I decided to go to Juan Fernandez, though to live like Selkirk or Crusoe I must do so on Selkirk Island, and I'm not quite sure how I am going to get there.'

'That shouldn't be any real problem,' said the officer. 'One of our government cargo ships will be calling into Crusoe Island within a few weeks, and I'm sure you'll be able to persuade the captain to take you the last ninety-six miles. It is a pity that our route is not that way. As it is, we are going several hundred miles off course just to go to Crusoe Island.'

'Yes, I know,' I said, 'and I appreciate it very much.'

'There is one thing I must ask,' another officer chipped in. 'Crusoe had a *Man* Friday. How come you have a *Girl* Friday?'

'How can you ask such a thing, Oscar?' said another officer. 'Wouldn't you rather have a girl than a man on an uninhabited island?'

'It's not like that . . .' I began, but was drowned out in raucous laughter, suggestive remarks and table-slapping for more wine. 'A toast to the second Robinson Crusoe,' one suddenly announced, and the six Armada officers sprang to their feet, clutched the table for support, braced themselves against the ship's roll and dip, and raised their glasses.

Hardly had they sat down again than another officer jumped up to propose another toast. 'To Girl Friday,' he cried. 'To Girl Friday,' they chorused. 'May Crusoe keep her happy.'

I had also stood. 'I'll drink to that,' I said.

'To illustrious *Esmeralda*!' Once more I drained my glass.

'To our glorious president – Pinochet!'

'God bless him,' I slurred. 'And all who sail in her – no, that's not right.' But no one was listening. Someone had thought of another toast. And so it went on. I have no idea who I toasted that evening. I do remember the waiter passing me a message that my Girl Friday would not be joining me for dinner.

Two

Esmeralda took twenty hours to cross the Humboldt – then she behaved like a serene, majestic lady once more. I located my Girl Friday and we sat on the warm deck planking, sunbathing in the lee of a large capstan. 'I wasn't actually sick,' she said, 'but I certainly felt groggy.' Cadets soon clustered around, the attraction being, of course, her dark curls, rose complexion and extremely well-filled sweater. A former social worker and nanny, twenty-one-year-old Ann Conroy was never short of admirers.

I hadn't fooled myself at all about the real reason why I had advertised for a Girl Friday – I was looking for another Rosemary. Alas, she didn't materialise, although I was again inundated with replies. When I selected a short list of three, my criteria were attractiveness, conversation and a lifestyle which would contrast sharply with a wild, uninhabited island existence and make a better story. While the other two women were more mature and had made it clear that an intimate relationship would be quite natural, Ann had declared 'no sex'. 'Of course not,' I said. 'We're British.' I liked the way she laughed at that. A major factor influencing my final choice was that the other two women were probably looking for husbands.

Platonic friendship was not a new experience for me – I had shared one for six years with Carol after the initial passion had gone. 'We have the best relationship there can be between a man and woman,' she would often say. 'In fact, you are my best friend.'

Carol's world was millionaires, trans-world trips on Concorde, weekend tête-à-têtes in Paris, Ascot and wild champagne parties. 'It's all false, Gerald,' she said. 'The only time I am the real me is here in this cottage with you.'

I lived with her for two months while waiting for replies from governments and I suppose I was living an island existence even then, so wrapped was I in my project. Every Friday afternoon she would be off somewhere, and every Sunday evening she would phone from Manchester airport just before getting into her 130mph BMW. I would have her gin and tonic ready for when she entered, looking like Zsa Zsa Gabor in her silks, furs and jewellery.

'Right, now I can get all this rubbish off and relax,' was her usual following comment and she would come back downstairs naked under her favourite caftan. We shared intimacy without being intimate. Magnetic vibes would often create frustration but we both knew we had reached a stage where to step across that platonic threshold would have ruined a fine relationship unless we went to the altar afterwards. Instead we enveloped ourselves in an unfalse aura of humour and gaiety which always ensured that any party we attended or held never went phut.

She is the only woman I have known who could sink gracefully to the carpet at a function and lie in a drunken stupor looking absolutely composed and beautiful. Always, she was surrounded by fly-bulging buckeroos who would regard me with open hatred as I'd help her to her car and secure her with the seat belt in the passenger seat. Then I'd push in our favourite Don Williams cassette and swish at a steady ninety miles per hour through the night with a beautiful, shapely bulk of liquid hanging in the seat belt next to me.

Once or twice she asked me: 'Why can't you stay here and write a bestseller about Cocos?'

'Because I haven't done what I set out to do. Until I do, I am incomplete and have nothing to offer anyone.'

'Balls,' she said, or something similar. 'Talk about a reluctant author. I think you're afraid to sit at that typewriter. Rosemary isn't. Look at her. She's had four books published.'

Many, many times, her friends would ask: 'When are you two getting married? You're so ideally suited.' Without exception they never truly believed that we weren't sleeping together.

The small, lobster-fishing community on Crusoe Island thought exactly the same about Ann and me when Forestry Commission chief Bernardo Ackermann allocated us a small wooden bungalow, free of charge, on the edge of a forest beneath the statuesque El Yunque (Anvil) Mountain where we lived for four weeks, waiting for a ship. They didn't believe we slept in separate rooms under five blankets each. They thought we'd sleep together, if only for warmth! It was August 1980 and winter in that part of the world, recognisable by extreme drops in temperature at nightfall.

For Ann, it was not a happy time. Three days after we were put ashore by long boat, she asked me to look down her throat. It was a mass of pus. She received a penicillin injection at the local clinic and was very ill for two days. Often afterwards I would notice her holding her throat and I began to worry. Septic tonsils are not the sort of things one takes to an uninhabited island. I was also worried about provisions and planted as many vegetable seeds as possible in boxes to take with us and not lose valuable growing time.

We had arrived at the island in the dead of night on the third morning of the voyage, the noise of the anchor chains awakening me. The portholes were black so I went back to sleep. At 7 a.m. I went up on deck for my first look at the place where Selkirk had lived alone, more than two-and-a-half centuries ago. God, it looked bleak. My first impression was high, barren rock and dark-brown cliff faces with very wet cloud rolling and curling over them. Small houses and buildings sprawled and huddled along the shore and I could just make out a tiny landing stage. We were anchored far out in the deep, cold-looking waters of the island's only harbour, Cumberland Bay. Ann joined me at the rail, zipping up her anorak. 'I though you said it was a tropical island,' she sniffed. 'I can't see a bongo tree anywhere.'

'I think some of the brochures we have are lying,' I said.

'This sea is nowhere near the claimed seventy-one degrees Fahrenheit. No wonder Selkirk wore goat skins.'

But two hours later, when the sun had melted the clouds, everything was brighter, different and warmer. The island was verdant with foliage. I called it the island of many lands and an awful monolithic testimonial to man's ignorance, stupidity and greed; a blueprint of man's destruction of the world; a multi-coloured lesson in anti-ecology which avaricious man will never heed. What a pity, I thought, that it is so remote from the rest of the world or could not be put in some vast museum in the centre of civilisation together with its history as an exigent example and warning.

When Spanish navigator Juan Fernandez discovered Mas a Tierra and Mas Afuera in 1574 they were covered with chonta palms and heavenly scented sandalwood. 'Ah,' he probably thought, 'here's a virgin, untouched place – let me see what I can do with it.' His choice was Mas a Tierra where one could trip over the world's largest lobsters, so numerous were they in the shallows, watch fish fight to throw themselves on the hook, plant all kinds of vegetables and fruits in the fertile soil, and clobber one or two of the hundreds of thousands of seals for a drop of cooking oil. He could build himself a splendid house and a new ship of sandalwood and elegant furniture with the chonta. So he lived there in paradise for five years. 'I need a nice piece of meat to vary my diet,' he reasoned, and introduced goats. Then he planted dozens of different food plants and, inevitably, weeds.

Everyone wants to see what's on the other side of the hill or mountain, so he blazed trails everywhere, exposing the ground to erosion. I had been guilty of the same thing on Cocos. Then Fernandez decided to go on to other things.

A year later, in 1580, a small Jesuit colony settled in Cumberland Bay, preferred pork to goat and introduced pigs. Sixteen years later, the colony was abandoned. By the time navigators Schouten and le Maire arrived in 1616, goats and pigs were running everywhere.

The next report is by a man called l'Hermite, who spent several days in Cumberland Bay in 1624 and, apparently, spread the word about the 'abundance of sandalwood and other trees for ship-building purposes'. He also saw quince, the stuff that the Owl and the Pussy-cat ate with a runcible spoon.

Forty years later, when a Father Diego de Rosales landed to start a colony, the cutting and exporting of the sandalwood and chonta were, like the axes, in full swing. It was the era of the buccaneers, and wood was almost as prizeworthy as gold and silver. Many seamen left their pirate ships for short spells to earn good money, helping to fell the timber. In 1686, the infamous Captain Davies's crew settled and were taken off by Captain Strong. Without doubt, Britain was the main culprit in denuding the two islands. The Spaniards were there, though, and to stop the British pirates from having such hearty feasts, put large, savage dogs ashore to kill off the goats and pigs.

By the time Selkirk landed in 1704 all colonists and lumberjacks had gone home, leaving the slopes bare with a few goats bleating. So even Selkirk, the prototypical Crusoe, never lived on an island where no other man had been.

For sixty-four years up until 1814, the island was a penitentiary. That same year the first of the present free colonists landed and, to put it bluntly, were an instant pain in the arse to both the Chilean Government and the island.

Fire constantly ravaged the island and, as a terrible indictment against man, by the middle of the century, no less than three million seals had been wantonly slaughtered. Also, like the sandalwood, the Juan Fernandez sea elephant was exterminated forever.

The indictment became worse and sharply indicative of the blatant, wilful destruction of nature today – for it was during that first half of the last century that botanist David Douglas and geologist Dr Scouler visited the island, and many naturalists followed. Yet never were any steps taken to stop the devastation.

A few cypress and eucalyptus trees were planted in 1900, but even when the island was made a national park on January 16, 1935, there was no real control. In 1948, five thousand sheep were adding to the island's destruction. Perhaps the real 'idiot award' should go to Desiderio Charpentier – he introduced *rubus ulmifolius* as 'living fences' to keep the sheep in! Sounds good – especially if one likes blackberries. What were once lush pastures are now hectares upon hectares of brambles. I couldn't ascertain who put the rabbits there, but those arch fornicators soon over-ran the island, almost in competition with the blackberries.

Three

EVEN WITHOUT being accorded the convenience of the entire contents of a well-placed shipwreck, as Defoe gave Crusoe, survival on an uninhabited island is really nothing heroic, provided one has sufficient provisions to last the three months that vegetables take to grow, and provided there is fresh water, game and fish. Indigenous edible fruits and food plants are of course an added bonus, and on such islands where they abound there is really no need to take provisions.

But trying to survive in a civilised, set island community, which has only a small shop, no gardens, has almost picked the shores clean and has imposed certain laws about killing land animals – especially those that are domesticated – is a different matter entirely.

'I don't understand these people, Ann,' I said. 'They grow nothing and rely on the government cargo boat for everything. When they are not fishing, all they do is drink and play sport.' One brochure called them existentialists, but I felt they wanted both sides of the coin – freedom from responsibility with government subsidy.

Ann and I needed protein other than that contained in our five kilogramme bag of hard black beans, so I immediately set about fishing. Our bungalow was about three-quarters of a mile from the village and a quarter of a mile from the sea. I walked along the beach and singled out a rock near a cove, sheltered from the constant wind blowing from the Antarctic, sat upon it in the hot sun and threw out my line, using crab as bait. Within a few minutes I saw a long, black shape gliding through the water. 'Oh, no! Not here as well,' I groaned.

But it wasn't a shark – it was a fully grown seal! She surfaced a few yards away and regarded me with round, brown eyes through long lashes that most women would give their eye-teeth for. '*I*,' I said, putting a harsh note into my voice, 'am supposed to be fishing.' I could have sworn she giggled. Rolling on to her back, she curved her tail over her belly, folded her flippers across her chest, gave a tremendous, ostentatious yawn, closed her eyes and went to sleep!

'Hey!' I called, softly. She turned her head and opened her eyes, questioning. 'Piss off!' Her answer was another wide, noisy yawn. Then she went back to sleep. I thought about throwing a stone at her but decided she had more right to be there than I had. 'I'll sit her out,' I said. Half an hour later, I gave up.

I told Bernardo about it on the fourth evening when he came to the bungalow to see how Ann was making out. He chuckled. 'She wanted company,' he said. 'Luckily, a few seals escaped the massacre and there is now a colony of about fifteen hundred on the south side of the island. It's not often they come to this part.'

Of German descent and proud of it, Bernardo had headed a six-year, uphill struggle to restore the indigenes to Crusoe Island. At fifty-six, he had reached the highest rank possible as a field officer for Conaf (Corporacion Nacional Forestal): 'Jefe Area, Experto Terreno', which meant he was chief of the region and knew everything about it. Besides being an arborealist, he was a veterinary surgeon, agriculturist and cowman. As a conservationist, he regarded the fisherfolk with a biased eye.

'But at least they are confined to Cumberland Bay,' I said. He nodded – reluctantly, I thought. 'What about tourists?' I asked.

'Not many come here,' he replied, 'which is good.'

'I hear there is a strong chance that a four-hundred-room hotel will be built on the shore,' I prompted.

'Yes, and that is very bad,' said Bernardo. 'This island should be solely for science. There are plenty of other places for tourists.' I heartily agreed.

Many times I returned empty-handed from fishing, simply because there were no fish there, no matter how many places I tried. But there are several things one can do with initiative instead of ingredients. I invented a favourite, filling dish which Ann and I entitled 'Curried FA'. She copied the recipe: take one medium-sized onion, chop and lightly fry in vegetable oil. Put a heaped tablespoon of ham and pea soup powder and ditto cream of chicken powder into a saucepan, pour on cold water and stir into a runny paste. Cook gently. Mix in three teaspoons of tomato puree, a pinch of oregano, salt and pepper, and a tablespoon of curry powder, and pour into frying pan. Mix with onion, adding more cold water if necessary. Simmer for twenty minutes. Serve with boiled rice or spaghetti.

Underneath, Ann wrote: 'Sufficient to make two people exceeding sick!' Alternatives were to sauté potatoes with the onion, or use pre-cooked hard beans or lentils instead of soup powder. The first time Ann tried cooking the last, she said, after we had eaten: 'I've got a new name for that – FF – Fucking Foul!' It was a real treat when I returned with a fish, especially the fierce-fighting jurel (yellow jack) which I was able to land after making a spear-like gaff. The jurel is a carnivore, but I was able to catch them, using crab legs as bait, by throwing the hook far out and trolling it back.

Because it became cold as soon as the sun dropped, Ann and I would sit with blankets round us in the evenings, reading, writing, listening to our several pieces of music on a battery cassette player, or talking about what we would do on Selkirk Island.

'I shall make a bikini of goatskins,' she said, dreamily. 'And I want you to take photos of me, standing on a rock with a knife between my teeth, and diving into the sea to get our dinner.'

'I see,' I replied, smiling, 'I suppose I've got to do the cooking while you play Diana.'

'You say I'm a rotten cook,' she pouted.

'You *were* a rotten cook,' I corrected her. 'You can even make bread now, and scale and gut fish. The meal you did this evening was absolutely delicious. These past two weeks have been good training for you.'

'I can just see our hut,' she went on. 'We'll have a roof of goatskins.'

'First, catch your goat,' I remarked.

'Ah, that's another picture I want you to take. You'll make me a beautiful bow and I want a photograph of me, driving the arrow to the feathers into a goat's heart.'

Bernardo had told us what Selkirk Island was like: 'It is more beautiful than here – warmer, too, because it is farther from the Humboldt. Living there will present little hardship to a man like you – more than a thousand wild goats range its twenty-one thousand acres. You'll find plenty of lobsters in the shallows, and the fish – ho! – the sea is full. You'll get a few sharks in the summer months, but you know all about those. There are quite a few fruit trees, fresh water all year round and the soil is very fertile.'

Several times, attempts had been made to colonise the island, the last inhabitants being taken off by the government the previous year. According to Bernardo, they had been a 'nuisance', calling on the radio for a ship whenever a minor illness occurred, or if they wanted to see relatives on the mainland. Their fires had also ravaged the island and they had paid little heed to fresh tree plantings in their hunting of the goats. There was no bay as such, but it was possible to land at one spot on the west shore where a few houses had been built.

'But the best spot for you is further south on the west side which has a fertile valley, full of sunshine and always sheltered from the wind,' Bernardo added. 'A river from the mountain runs through it to the sea.'

'It sounds like paradise,' I said.

When Bernardo had gone, Ann clenched her fists and shook them with frustrated enthusiasm. 'I can't wait to get there,' she said. 'Just the two of us and no one else around. If only that ship would come! I want to prove to myself and to the world that I can do it. I'll make you a damn good Girl Friday.'

It was during our third week there that Bernardo arrived one evening with the old Nazi concentration camp joke: 'I have some good news – and some bad news.' The good

130

news was that a ship called the *Aquiles (Achilles)* was arriving the following week on September 10th. The bad news – Argentina was threatening war over the ownership of Tierra del Fuego, all Armada ships were on stand-by and the *Aquiles* was returning immediately to Valparaiso.

'We have been told that it will be November – more likely December – before another ship comes,' said Bernardo. 'That is why the *Aquiles* is being used – she is a big troopship and can carry lots of cargo.'

'But surely there must be some way we can get to Selkirk,' I demanded in fear-filled desperation. An uneasy premonition was growing within me that this venture was to be another abortive attempt to be Robinson Crusoe. Just when everything seemed to be going well, too. I cursed the Argentinians.

'None of the boats here will travel that distance at this time of the year, Gerald,' said Bernardo. 'In the summer, yes. But not now. There can be waves thirty feet high out there.'

'What are we going to do?' Ann asked, after Bernardo had gone.

'I really don't know – wait, I suppose. But if the next ship won't take us, it's going to be pretty grim waiting until January or February for one of the fishermen to give us a lift.'

'Grim!' she exclaimed. 'It will be awful! I don't like this island or the people. Half of the men behave like barbarians.'

I knew what she meant. One of them had demonstrated what a hunk he was by slitting a sheep's throat in the small field in which the bungalow stood; then he'd put his lips to the spurting blood, pounded the dying sheep's ribs with his knee to get the last drop, leered at Ann with blood-smeared face, thumping his barrel-chest and declaring: *'Haceme fuerte'* (Makes me strong).

Also not very endearing to a sensitive, twenty-one-year-old woman was the way they poached the few remaining wild goats on the island after Bernardo had confiscated all their rifles to stop them from exterminating the goats completely. They tied several large fish hooks on to the end of a rope which they threw at the fleeing goat.

'Oh, the thought of fish hooks going into their flesh gives me the creeps,' she said.

'They are a strange breed of men here,' I said. 'They don't associate pain with animals. I doubt if there is one dog on this island that would dare to bite anyone. It knows that if it did, it would be dead instantly. The dogs are not obedient because of joy in pleasing their masters – they obey because they are in what I term subdued subservience inculcated with a boot.'

There was another thing bothering Ann. She had told her parents that she would be away for only a year. The longer we were on Crusoe Island, the less time there would be on Selkirk. More important, in the light of the present situation, what chance was there that we would ever be taken off Selkirk if we got to it?

During the week before the *Aquiles* arrived, Ann became very restless, listless and rather short-tempered with me. She complained more about not being able to sleep at night because of the strange noises of the wind. Even in the daytime it could be quite alarming, especially if one were in an isolated spot. I likened the wind to the sound of an approaching train on London's Underground – a faint rumble in the distance, getting louder and louder. Then the wind, being forced through the tunnel, suddenly struck, and the train clattered in, vibrating the platform. On Crusoe Island, one could hear the approaching wind through the trees, increasing in volume. When it struck, the noise was deafening, the branches and trees bent and some of the leaves went flying. Then all was still – and one found oneself waiting for the next train to come along in about thirty seconds.

At night-time, the wind moaned and groaned, moving the bungalow with creaks, sighs and whispers. Often I heard what sounded like voices and footsteps in the living-room. On the first occasion I asked Ann if she had got up in the night. 'No,' she said, 'but I thought I heard you moving about.'

Two days before the ship arrived, she complained that her throat was hurting again. Her tonsils and back of throat were inflamed. 'I'll ask the captain to let the ship's doctor take a look at you,' I said.

'I don't need you to ask for me,' she flared. 'I'm quite able to speak for myself.'

The doctor immediately sent a longboat to shore for Ann's belongings, confined her to sick-bay and told me he was placing her in Valparaiso hospital for observation.

'What's wrong with her?' I asked.

'I don't know,' he said, 'but we'll find out. Don't worry about money. The Armada will pay and deliver her back safely to you even if we have to put her on one of the small planes that come in to the island.'

We discussed something I had been interested in for many years – self-induced illness. 'It could be psychosomatic, but I don't think it is,' said the doctor. 'By the way, have you got your appendix?'

I knew why he asked. 'Yes,' I said.

'I think I had better show you how to perform an appendectomy. You've got nothing to lose, having a go at it yourself because you'll die in any case if you don't operate. It's quite a simple operation, really, and you'll be in such agony that you'll want to cut it out to stop the pain in the way you crave for an aching tooth to be pulled.'

He drew me a map, showing me exactly where it lay and made a parcel of Savlon liquid, swabs, sutures and two scalpels. I always regarded the package as a cyanide pill and often wondered what it would be like to play the great doctor Frobisher.

'What about an anaesthetic?' I asked.

'You won't need that,' he said. 'The pain will anaesthetise the incision. Teeth OK?'

I told him I never had trouble with them. 'They're all mine except these two front ones which a boxing glove took away when I was 19.'

An old cliché, but the bungalow was empty without Ann. I took photographs of her bedroom walls adorned with her poems and sketches. What sadness and loneliness they depicted. Bravery, too. Had I been an inadequate companion, guilty of indifference again and too ready to criticise ineptitude? An afternoon in the small rowboat Bernardo lent me came readily to mind as an example. The day before, I had rowed across the deep waters of the bay, filled with romantic thoughts of Selkirk and pirates. Always, I have been able to immerse myself in a place's historical aura. Sometimes that presence could become over-powering as it had when I stood in the prison caves in Rome's Coliseum and could actually hear the cries and screams of the Christians, the roars of lions and the yells of the crowd. I rowed over the spot where the German battleship *Dresden* lay in forty fathoms, and easily imagined the scene in 1915 when ships of the Royal Navy penned her in, forcing her to scuttle rather than surrender. A shrine to her crew stood in the small cemetery on the shore.

Ann accompanied me the following day, stripped to her bra when we were a distance off-shore and, despite my warnings, burned her back. Then I had to interrupt my fishing and put her on a rock because she felt dreadfully seasick. I made a few scathing remarks – not solely because of her sunburn and seasickness but because she was unaffected by the romance of the island's history, and appeared disinterested in Selkirk, the man. That evening I received a minor ticking off from Bernardo. Several people had watched me through binoculars – or was it Ann they had focussed on? – and complained to Bernardo that I had 'flagrantly disregarded' harbour safety regulations by going out so far in such a tiny boat.

During the first week on my own, I retraced Selkirk's footsteps and visited the two caves – one on the north side and one on the south – which have long been the subject of arguments over where he actually lived. I climbed the two thousand seven hundred metre trail to his look-out point, five hundred and fifty metres high in a saddle on a razor-edged ridge where views of the South Pacific on both sides could be commanded, and where it felt like the wind would take my head off. I was also accorded a magnificent view of the archipelago's third island, Santa Clara, lying off the southern point and resembling an island in the Blue Aegean. Little more than barren rock, Santa Clara has sufficient vegetation for a few wild goats and was once called Goat Island.

Two plaques to Selkirk had been placed on his look-out. The larger, of wood and erected by Conaf, declared in rather verbose and poetic Spanish: 'In this place day after day during more than four years, a Scottish sailor, Alexander Selkirk, anxiously scanned the horizon in search and hope of a rescue ship to liberate him from his isolation and allow him to abandon his solitary domain in order to return some day to his similar kind and maybe return to his native land.'

The second, of iron and cemented on the side of a rock, was in English: 'In memory of Alexander Selkirk, mariner. A native of Largo, in the county of Fife, Scotland. Who lived on this island in complete solitude for four years and four months. He was landed from the *Cinque Ports* galley, 96 tons, 16 guns, AD 1704, and was taken off in the *Duke*, privateer, 12th February, 1709. He died, Lieutenant of HMS *Weymouth*, AD 1723, aged 47 years. This tablet is erected near Selkirk's Lookout by Commodore Powell and the officers of HMS *Topaze*, AD 1868.'

According to my research, Selkirk died in 1721, his epitaph being four words in the ship's log: 'PM. Alexander Selkirk. Deceased'. That evening I wrote in my diary: 'If his passing went unsung at the time – even to the extent of making a mistake about his year of death – I will do all I can to ensure that he is better remembered as the man who unknowingly sat for Daniel Defoe's undying portrait of Robinson Crusoe who gave people of this world an oh, so necessary romantic escapist dream.'

For a start, I typed, immaculately, the Spanish translation of the commemorative tablet on a sheet of pristine white paper, substituting 1721 for 1723. Bernardo framed it and hung it on the wall in Conaf's small museum near the landing stage.

The southern part of Crusoe Island lies in a rain-shadow and is little more than mesa and desert, tapering to a peninsula. One late afternoon when I had decided to spend the night in the dank, southern cave, I walked through the centre of the mesa and was startled by the thud of galloping hooves. Racing towards me was a herd of horses. I stood still, taking in the liquid movement of their chest muscles, flared nostrils and flying manes and tails. Suddenly, they veered, heading for the high grasslands where the rain-shadow ended.

I proceeded on down to the narrow desert of white and yellow sand for a southern view of the island's backbone range of mountains with El Yunque standing aloof in the middle. The setting sun had turned the peaks to fire and I had the impression of sharp-pointed flames rising from hell. It was when the sun sat like a giant red ball on the horizon that Selkirk's Lookout accorded a glimpse of the tip of Selkirk Island's five thousand five hundred foot mountain to the right of the setting sun. At no other time was it visible to the naked eye. On that mountain lived Bernardo's proposed answer to the rabbit problem – a large, white falcon found nowhere else in the world. 'I am hoping to go there in January and trap a few young ones,' he had told me. 'Then I'll release them here.'

Often I would stand alone on Crusoe Island's various shores at night looking at the needle-peaks and savage, saw-tooth outline of the central mountain range against the stars and contemplate those awful, terror-filled first eight months that Selkirk had experienced.

His terror was not of starvation or ghosts but of loneliness. So frightened was he of missing a passing ship that during those months he never left the beach. With no one else to turn to, he sought God and read his Bible daily. Slowly, he became reconciled to his fate and built himself a hut near the edge of some trees. The island itself was pleasant, the climate healthy and there was an abundance of food. After a time, he moved into a small cave and blocked the entrance with stones, leaving only a small hole to crawl in and out through. A dilemma was that he might be spotted by a Spanish ship and taken prisoner or killed.

For food, he hunted the wild goats and ate fish until peace of mind led him to vary his diet and to try vegetables and fruits. He enjoyed adding things to his shelter to make his life more comfortable. To ease his solitude, which he found himself beginning to enjoy, he caught and tamed some of the wild cats on the island. When his shot and powder ran out, he became so fleet of foot that he could out-run a goat and kill it with his knife.

I found myself enjoying my own company and studied my book of quotations to see what past masters had written about loneliness. My favourites were: 'So lonely 'twas that God himself, Scarce seemed there to be' (Samuel Coleridge, 'The Rime of the Ancient Mariner'); 'O Solitude! Where are the charms, That sages have seen in they face? Better dwell in the midst of alarms than reign in this horrible place' (William Cowper's interpretation of Selkirk's thoughts); 'Whoever is delighted in solitude is either a wild beast or a god' (Francis Bacon); and 'No man is an Island, entire of itself; every man is a piece of the Continent, a part of the main' (John Donne).

It was during the second week on my own that I found I had a companion – a wild cat. Having found the best fishing spots, I was eating quite well, and a large tabby cat was frequently raiding my refuse pit. I left morsels on a plate outside the back door and, over the weeks, 'Cat', as I called him, grew less timid and would eat in the kitchen.

Eventually, 'Cat' would sit next to me in the evenings. Finally, he allowed me to stroke him, evoking a rasping purr. We did have a minor setback to our growing friendship early on when he stole my fish and I hit him with the dishcloth. Instead of fleeing, he growled and struck at my trouser-leg with bared claws. We glowered at each other than tacitly agreed to forget the incident.

It was, most likely, my taming of him which caused his death. One fine spring evening when we were eating our fish outside, with my favourite hummingbird gleaning the nearby fir of insects, a large Alsatian mongrel stepped round the corner of the bungalow and almost tripped over 'Cat'. The surprised dog snapped instantly and instinctively, breaking the cat's backbone. The dog fled with clods of earth falling around its tucked-in tail.

I tapped 'Cat' behind the head with the axe and buried him under a eucalyptus tree, placing a simple wooden cross at his head. My diary entry that evening: 'Melancholy – what is it? How do others define it? Nothing explanatory. I think my real trouble is that I am always angry at the futility of life. For instance, why did Horeb have to die? And why did I break my heartfelt vow? Happiness – grief – happiness – grief. That's all life is.'

Thoughts of Horeb flooded in, diffusing and surpassing the demise of 'Cat'. It is said that a sailor knows only one ship that he loves, and that there is only one man for one woman. I am sure the same is true of a dog. Horeb was a magnificent red setter who loved me to everything and everyone else's exclusion. He fretted when I wasn't in sight and even knew when I winked at him. He was as stupid as Pluto, and I remember one day the boys and I collapsed with mirth after a rabbit ran through his legs and he stood in ridiculous pose with head between front legs to see where it had gone.

When I took him to one of Italy's leading vets I said there was no need to muzzle him while I was there. Horeb sat with front paw raised, trust and devotion in his eyes, while the vet made three jabs before finding a vein. Horeb's attitude was a bored 'What's this arsehole doing?' The vet diagnosed pernicious anaemia and cancer of the colon and called to put Horeb down on the day I was in Rome on urgent business. I arrived back to a house of weeping. With red eyes and tears streaming, Rosemary told me that the vet had laid him on the stable floor. I switched on the stable lights and sat on the straw next to Horeb, pretending he was sleeping and fighting back the tears. Rosemary came in twenty minutes later and we held each other and sobbed.

Next morning I buried Horeb near Napoleon's sister's

138

favourite rustic seat, overlooking the heat-hazed Mediterranean and the lake and house where Puccini wrote 'Madame Butterfly'. Puccini's inspiration had been the delicate sweep of the Tuscan hills where Horeb, at the age of four, would lie forever. 'I'll never, never leave you alone in this foreign land,' I vowed over the grave. But I did – and always felt a certain guilt and inadequacy for doing so.

Two weeks to the day after Ann was taken away, Bernardo brought me a telegram. No diagnosis had been made and she was returning to England.

'What will you do?' Bernardo asked.

'Go on alone if I can get there,' I said.

A week after I buried 'Cat' and after nearly six weeks of living alone, a mixed delegation of teenagers paid me a shy visit. The spokesman, frequently elbowed by the others, told me I was invited to an annual festival next day.

The music, dancing and drinking went on for three days and nights. I collapsed, legless, on the second. The fisherfolk were extraordinarily friendly people. Invitations to dinners at their houses and to go on fishing trips followed, and my Spanish improved considerably. I was drawn into community life.

Bernardo took me in the Conaf boat all round the island and we invaded the seal colony. A huge patriarch rose out of the water close to us, looking and bellowing like King Kong. 'His bark is worse than his bite,' said Bernardo. I loved the mischievous faces and pranks of the cubs, and wondered how any man could bring himself to club one.

One day, when we were passing through the strait between Santa Clara and Crusoe Island, parts of the sea were bright red. 'Lobster spawn,' Bernardo explained. 'A female lays between sixty and eighty thousand eggs, but only ten per cent live and survive to become mature lobsters. With the amount of fishing now taking place, it is estimated that the yearly potential loss to these waters is two hundred and ten million. In another five or ten years there will be no lobsters here – then the fisherfolk will be in trouble. They'll have no income.'

'I know,' I said. 'The Armada officers told me about that problem. When I asked what the government would do, one of them said "Probably shoot them!" They want to be islanders, apart from Chile, yet constantly call for government aid. A man could live in true existentialism here. Why don't they just become self-sufficient islanders?'

Bernardo rubbed his thumb and fingers together. 'Where would they get the money for their expensive clothes and radios? And where would they get their wine?'

'Grow it here,' I said. 'This island would be ideal for vineyards.'

Bernardo shook his head. 'Anything that involves work, they shun.'

On another day, Bernardo took me by mule to many parts of the forty-square-mile island, and I found it to be another place of many lands. The High Wastelands in the north-east, with their cattle grazing in the valleys, could have been carved from Colorado, for their soil and rock faces were surely red enough; with Conaf's re-plantings, parts of Cumberland Bay reminded me of Switzerland or Scotland, the scent of pine heavy in the air; over a hill, and Tuscany came readily to mind, the thick clusters of eucalyptus leaves silver-green like those of the olive. Then there was the Grecian isle appearance of Santa Clara, backed by a portion of the High Sierras from Spain. Up there, we came to a rock and shale ridge, hardly more than a yard wide and sheering two thousand feet on either side. 'Throw the reins on his neck,' Bernardo called to me. 'He'll take you across.' I did so, and the mule never missed a step.

The fisherfolk make their own nine-metre open boats of eucalyptus ribbing and cypress planking. An outboard motor, placed inboard, drives them at a maximum seven knots and makes heavy going in the winds and rough seas outside Cumberland Bay. Yet, almost every day the sky was azure and the sun very hot. Once when we were battling the wind and waves across a sweeping, inward curve of cliff face on the west side, Bernardo told me that thousands of feet below was a graveyard of Spanish galleons. 'They were all heavily laden with gold and silver,' he said. I told him that the area reminded me of a miniature Bay of Biscay.

Four

BERNARDO'S WIFE, Rachel, had arrived on the *Aquiles*, and I'd noticed that he'd put his twenty-six-year-old house-keeper, Anna, on board with Ann. The night before the ship arrived, Bernardo came to the bungalow to borrow my powerful torch. With his blue eyes crinkling, he said: 'People will be coming ashore about two o'clock in the morning. My wife is enormous, very heavy and she can't swim. If she trips getting into the boat she'll go to the bottom like a rock! I'll need a good torch to search for her through the depths.'

I appreciated that Bernardo had a sense of humour when I saw Rachel. The antithesis of Bernardo's description, she was a petite, trim-waisted and raven-haired woman in her forties. It wasn't until after the festival that I got to know Bernardo and her very well, and they both laughed when I mentioned the incident.

Rachel's problem was Anna, who was obviously more than Bernardo's housekeeper. Rachel had shared little in Bernardo's six years on the island and was not at all happy. 'I have stayed on the mainland to see our two children through university,' she told me. Often she would prompt me, even in front of Bernardo, with: 'Anna is very beauti-ful, don't you think?' I always replied that Anna was only twenty-six and not my type. In return for meals at their house, I made wine from quince and *nespero*, and Rachel tippled daily.

In the after-dinner discussions with Bernardo, who always liked to interject a German word or two into his Spanish, I learned he was dedicated to his work on Crusoe Island and devoutly believed in, and bewilderingly re-vered, the wonder and power of life within seed. 'My one regret,' he said, 'is that I shall be dead when this island is

142

covered once more with the mature chonta palms I am growing from the seeds of the few survivors. My constant wish was that I could bring back the sandalwood.'

'Is there no sandalwood left anywhere in the archipelago?' I asked.

He shook his head. 'All gone. I have searched all the islands, but there is nothing.'

'Surely there's sandalwood left somewhere in the world,' I said.

'There is,' he replied, 'in India. But that is no good to me. It is not indigenous to Juan Fernandez.'

We touched upon religion and I told him how I had tried to imitate Selkirk and Crusoe by reading the Bible in the evenings in the bungalow. 'I'm afraid it still seems the mumbo-jumbo it has always seemed since I was thirteen. My grandfather was a parson, but I just couldn't believe and turned from the gospels of my own volition.'

'So you don't believe in God?'

'God knows what I believe in,' I said, and we both laughed. He fetched a few melon seeds and held them in his open palm. 'In each of these are generations upon generations of melons – sufficient to cover and strangle the entire world. And all we can see are seeds. You are not going to tell me that the wonder of life was an accident and not something that was planned and devised by a superior being.'

'Why not? It is only our ego that seeks a reason and the insatiable quest to determine it. Personally, I don't think we are of any more importance in this world than a rock. Astronauts see the world as a ball. They can't see all the complexities and anxieties and activities that are taking place on it, from the self-importance of dictators to the bigotry of business bosses or the urgency of a typist's staying late in an almost death-before-dishonour attempt to finish a document or letter; they can't see the traffic jams, the frustrated fuming of drivers whose individual journey is the only one that is essential – all the astronauts see is a smooth-sided ball. All we can see is a smooth-sided seed.'

143

'Yes, I see what you are getting at,' he said, excitedly. 'You are alluding to the two extremities of infinity. But can't you see, that's the whole point – it's too vast to have been an accident.'

'The symmetry of a snow flake is accidental,' I said. 'When I was a boy I was told it was the result of God's infinite care. Now I know that it is designed solely by vibration as it floats to earth.'

So our discussions would go on, encompassing atheism, agnosticism and many other isms. And always, as is the case with religious debates, no answer was ever arrived at.

The six hundred or so Islanders' weekend religions were netball and football. Every Saturday and Sunday, Cumberland Bay resounded with cheers, shouts and referees' whistles as the segregated teams of men and women representing the four family clans competed with fanatical verve, kicked shins and many a punch-up. Christian religion had been ousted. Even the pastor had left. On the one occasion I attended the small chapel on a lone part of the shore, a layman took the service. There were only eight people present. One of them was a striking blonde woman of thirty-two. Her name was Sandra Cruz and I found her personality overwhelming. I went home with her for drinks and she showed me the essence of her life contained in newspaper reports of her singing in opera all over Europe. She was the leading light of the community's operatic and dramatic society and, a week later, I saw her on the stage in the tiny, packed theatre. She captivated the play and held everybody spellbound.

'Her vitality is marvellous,' I commented to Bernardo.

'She is dying of leukaemia,' he told me. 'She knows it.'

I was thunderstruck – and contemplated once more what a bastard life could be and that it was absolutely pointless praying to Him, whatever or whoever he, she or it might be. He'd have a lot of answering to do if I ever met Him. I recalled that I had probably been the only British soldier in Korea who had refused to wear an issued rosary. I had asked nothing of Him before and I was damned if I would ask something of Him then.

Five theologians, each an expert in one of the five parts that the Bible comprises, had tried for four weeks, on an Army Christian Leadership course in the Hartz Mountains in Germany, to get me to see the light. And failed. I often felt – and still feel – it was a pity that man didn't keep to Christianity instead of being side-tracked by the church and papal power. During my days as district reporter on provincial newspapers I met many a recalcitrant clergyman who thought the same. Goodwill to fellow men was a simple, understandable, logical indoctrination to keep the peasants happy – why complicate matters with the indefinable, intangible but master-minded gimmick of faith in something that changed with each scientific discovery? The answer, of course, is that faith is the only salvation of man's inadequacy.

A vicar in Kent, whose company I frequently enjoyed, was a classic example. All the men in the battalion he commanded were killed at the Battle of the Somme. He stood alone, surveying the carnage, and he was overwhelmed with fear, helplessness and hopelessness. 'There was no secular being I could turn to,' he said. 'There was only God to help in my inadequacy, and when the war was over I took the cloth, determined to bring strength to the weak by showing them God as I had found Him.'

A memorable glimpse of distinct life and young love on Crusoe Island was accorded me one sunny Sunday as I sunbathed on the verandah. A slim young woman in a pretty print dress walked in the meadow below the bungalow and laid down in the spring greenness. Suddenly, a white charger sped towards her and she sat up with arms outstretched to greet the rider. He threw the reins over the horse's head and sprang from the ornate saddle while the horse still cantered. The young man went into her arms for a long kiss; then he walked back to the grazing horse, swung into the saddle and trotted back to the girl who now stood, waiting. Using his stirrup and arm, she mounted astride behind and clutched him round the midriff. Then, with her hair flying and dress thrown up around her thighs, she abandoned herself to the motion of the galloping horse and waved to me as they thundered past and disappeared through the forest.

What a picture of how lustful swains once courted maidens; and how much more romantic the horse than civilisation's equivalent, the sports car. The era of the horse had not been superseded by carbon monoxide on Crusoe Island. There was only one vehicle – a dilapidated, council jeep which went only when it felt like it. The houses were wooden and the streets reminded me of pictures of the old Wild West and the Yukon – mud when it rained and duckboards to cross them. Fireplaces were not allowed inside houses and when gas cylinders ran out, the women cooked outside on wood fires.

Household electricity was generated for a few hours each evening by communal diesel engines. Conaf's one had broken down, so I put my Army training in electronics and mechanics to use and repaired it, together with a petrol chainsaw. In return I was given extra provisions; and more food came my way by trading my wine recipes.

I suppose, because of the fisherfolk's idleness in planting nothing in the extremely fertile soil, life was a survival existence for many if a misjudgement had been made about the purchase of vegetables and provisions from the cargo boat, or if the boat arrived late. The shop usually ran out of perishable goods after two weeks. Households which couldn't afford the extortionate air fares charged by the small aeroplane company for passengers and freight simply ate fish with nothing else for every meal.

In my leisure moments I made an excellent spear for fishing and a cypress bow which astonished some local children and me by sending an arrow for more than two hundred and fifty metres. I was carving an experimental bow of eucalyptus when Bernardo brought me the worse news of all: the *Aquiles* was arriving in a few days' time and leaving almost immediately for Talcahuna, southern Chile. War was more imminent than ever and reservists were being recalled.

'The only way you will get to Selkirk is when I go in January,' he said. 'Disturbing news for me is that several fishing families have been given permission to stay on Selkirk until April or May for the lobsters.'

Disturbing it may have been for Bernardo. It was absolutely devastating for me. Anger gave way to resignation. 'All these weeks here for nothing,' I said.

'Why?' Bernardo asked.

'How the devil can I be Robinson Crusoe if there are going to be people on Selkirk?'

'Robinson Crusoe?' Bernardo exclaimed. 'I am Robinson Crusoe! I have been isolated in this community for six years! Stay here and help me. You can go across to Selkirk, build a hut and take a few photographs and pretend you have lived there. We are so remote from the world that no one would know the difference. And you should see the bikinis on this beach in the summertime!'

'I'm sorry, Bernardo, I couldn't do that. My book would be a lie, and that wouldn't do me or my conscience any good at all.'

Several times he asked me to stay and he told me that Rachel was leaving and Anna would be on the *Aquiles*. My vegetables in the garden I had made at the rear of the bungalow were as good as ready. Bernardo said he would think of me when he ate them. He gave me a tremendous bear hug when we said goodbye, and tears streamed down his cheeks when he kissed Rachel. But he had made his choice – Anna.

I had an awkward moment or two with Security at Talcahuna and was grateful for Rachel's presence. I was told to report to the nearest CID station for documentation or I would have trouble leaving the country. 'You'd better come home with me to Temuco,' said Rachel, 'and we'll get it all sorted out there.'

Her lovely twenty-year-old daughter Jeannie took charge of me, sorted out my papers with the police and helped me arrange my flight to England via Bolivia and the States. We became very close and I stayed there for two weeks longer than the few days I intended.

'Why don't you stay here and teach English to university students?' she suggested. 'When I leave university next spring I would go with you to Selkirk Island.'

'I can't stay here that long,' I told her. 'I am already long overdue getting this book done. I must go back to England and find another island – Selkirk is no good if the fisherfolk are going back there. Besides, I am thirty years your senior.'

'You are so old-fashioned,' she said. 'There is the same age difference between Anna and my father, and they have been very happy for six years.'

On our last evening together, I caught her writing something in my notebook. She sat, very pink of face, as I read it: 'All the promises of my love will go with you. Why go?'

Over the next years I had great cause to regret that I did.

Part Three

Tuin Island

10 10'S, 142 E

Torres Strait, between Australia and
Papua New Guinea

Badu Island

Three kilometres north of Tuin

One

IT WAS late May 1981.

The old, black-hulled cargo boat, so heavily laden and low in the water that her chugging diesel engine was hard-put to produce six knots, was in all likelihood following the same circuitous deep-water channel between reefs and around under-surface rocks that the great story writer, Somerset Maugham, had sailed several decades before. Any deep-draught vessel yawing from the charted course in the treacherous sea of the island-strewn Torres Strait would be soon holed; and there would be few survivors, for the warm, turquoise waters are the home of the vicious, twenty-foot tiger shark and – even more feared by the Islanders – the man-eating sea crocodile.

Our place of embarkation, Thursday Island, set within a cluster of other islands, had still smacked of Colonial days with its white masters and black subservients, and the steamy, mosquito-droning region readily reminded me of those remote equatorial reaches that Maugham frequently used as story settings.

Apart from roads, mechanisation and building technology on TI – as Thursday Island is known – and a small airstrip on nearby Horn Island, the island group had changed very little since Bible-clutching missionaries first walked among the cannibals and head-hunters, and the white Australian exploiters had rid the Strait of barbaric tribes with more forceful means – bullets. Whole clans, including women and children, had been massacred.

Torres Strait history makes good material for any blood-thirsty writer.

Every fortnight or so, the aptly named cargo boat, *Torres Strait Islander*, started her sluggish, hundred-and-thirty-mile round trip from TI to the midway islands of the western sector which, with others and the TI group, resemble giant stepping stones across the hundred-mile stretch of shallow water, separating Papua New Guinea from Australia's northernmost tip, Cape York.

Somewhere in that middle-distance group, appearing to me in the early-afternoon sun as a far-off, heat-dazzled blob of blue contours, was the tiny uninhabited island allocated to me for my third Robinson Crusoe bid. Its name was Tuin, which means 'garden' in Islander language.

Before darkness, we were close enough to see one of the two largest islands of the group, called Moa, in detail. Most of its interior was hidden by long, thick stretches of mangrove, and its mountains and coconut palms rose in relief against the reddening sky.

Tourism had not invaded this part of the world, and I had been told that all middle-distance islands in the Strait represented one of the few remaining 'civilised' places that were still unspoilt. The romantic within me reasoned that the island looked today as it had done when Torres, Cook and Bligh had sailed past, and I felt an immense, warm pleasure that I was gazing upon the very same picture that those great explorers had seen.

By mid-morning of the following day, after lying off-shore at anchor throughout the night, cargo had been delivered by dinghies to the two partially-hidden villages on Moa – each inhabited by little more than a hundred Islanders – and we were making our way to the other large island, Badu, where the village population totalled some two hundred and fifty.

With sweat streaming from my every pore, I stood in the comparative cool of the bridge next to Captain Bert, whose skin, like that of his crew's, was the colour of ebony, and received my first convincing warning about the Torres sun.

'I have been away for several months working in the south,' he said. 'I have only recently returned, and my back peeled after taking my shirt off for an hour or so!'

In amazement, I said: 'I never knew that black people's skins peeled.'

'I can assure you, mine did,' he replied. 'Be very careful.'

From the little information I had been given, I knew I wasn't going to a paradise island in any sense of the word. Small, deadly and hard-to-see death adders were on most of the islands. Badu, for instance, was covered with them. Then there were poisonous black snakes and highly-toxic redback spiders. Goannas – giant lizards, reaching lengths up to seven feet – were on all islands and the bite of one species could prove lethal. Non-poisonous pythons and green whipsnakes were plentiful.

In the hot shallows along the shores lurked the hideous and poisonous-spiked stonefish that could easily kill a man if stepped upon, and its companions – slightly less evil and repulsive – were two kinds of poisonous ray, one blue and one brown. Adding good measure were neuro-toxic sea snakes.

As though all those, plus sharks and crocodiles, were insufficient to give anyone just cause for dying, there was always the threat of drought and the fact that the sea itself was poisonous! The slime on the coral turned any skin abrasion into a painful tropical ulcer, and I had read several reports of deaths from coral cuts.

Yet I felt quite at ease and relaxed for, despite the dangers, I was at last going to an uninhabited island. No matter that I was penniless and tobacco-less, I would soon be in an element I had grown to love – island life and the complete, lazy freedom that went with it, where a man was rid of time and civilised programming, could be a boy again and do exactly as he wished – always provided, of course, that he was not accompanied by an over-enthusiastic and energetic Girl Friday called Lucy!

When Captain Bert first directed my gaze to Tuin, I did not call Lucy to the bridge for I wanted to savour the moment alone. With my first sighting of that palm-swaying, mangrove-swamped island, with its long, eastern stretch of sand the colour of gold, I suppressed a strong surge of excitement within me. With my two previous failed attempts, I resolved not to be too optimistic – though

153

never pessimistic – until I had been on Tuin for three months, for that was the telling time when food from the vegetable seeds I would immediately plant would take over from our meagre stores and make us truly self-sufficient. My skill as a fisherman would always ensure we had plenty of protein, for the Torres Strait waters teemed with edible fish.

There was another female on board – Jackie Mott, a vivacious, dark-curled photographer, commissioned by the London *Sunday Telegraph* and flown from Sydney to capture the final stage of our journey and the first moments on Tuin.

She and Lucy, who, with her severely-chignoned hair, long, mock-Victorian dress and sweat-streaked face, appeared as a gaunt lady missionary against Jackie's roundness, each held a camera and, standing for'ard, were firing at everything in sight. I went down to join them. 'You see that long island ahead,' I said. 'That's Tuin.'

Both cameras were instantly aimed and began clicking furiously.

'What do you think of it?' I interrupted.

'It's beautiful, dah-ling!' Lucy gushed in her inimitable fashion. 'Absolutely beautiful! From what we'd heard, I never dreamed it was going to be anywhere near so big.'

'Just think of it,' I said. 'It's all ours.' Then, with all the ego of a pioneer, I added: 'You see that big white rock in the centre of the bay? I'm going to name it "Kingsland Rock".' And Lucy agreed wholeheartedly.

Like me, she wanted to be put ashore by dinghy immediately, but courtesy decreed that first we had to go to Badu to introduce ourselves to the council chairman, Crossfield Ahmat, under whose jurisdiction we would be and who had granted us the ultimate and privileged permission to occupy Tuin.

As we rounded the island's northern end, I could see that it was very narrow in the centre; later, I was to discover that Tuin was shaped rather like a bone with flared ends, each of which rose to small, densely-foliated hills. All the trees and bushes were permanently slanted

154

in their growth because of the strong, incessant south-easterly Trades. The distance across Tuin's waist – or should I spell that waste? – was in the region of two hundred and fifty to three hundred metres.

Three or four miles further on, passing by two smaller circular islands, we reached the village of Badu and were taken ashore by dinghy. There, on the hot and palm-lined beach, with its casual but not intense untidiness of discarded petrol and beer cans, I met Crossfield. A tall, dignified, brown-skinned man of Malayan descent, he was five years my senior. As our eyes met and we shook hands, there was an instant feeling of liking between us. Here, I thought, was a man who would never let you down.

In the shade of a sparsely-limbed tree, he knelt to draw me a map of Tuin – which, indeed, did resemble a bone – to show me the position of the only fresh-water creek next to a garden shed, constructed of corrugated iron. The creek and shed were on the western side of the island which we had not yet seen.

Just before I got into the cargo boat's waiting dinghy to take us to Tuin, Crossfield once more shook my hand. 'I hope your womans good womans,' he said. 'You plenty responsibility for me. You get sick, you light big, smoky, signal fire. I send dinghy to get you out.'

'Don't worry about us,' I assured him, 'we'll be OK. Thank you very much for allowing us to live on Tuin and for your help.' It wasn't until many months later that I appreciated his way of pluralising the word woman!

The fifteen h.p. outboard motor should have sent the twelve-foot aluminium dinghy skimming across the water at about twenty knots, but not with the load we had on board: two cargo-boat crew, Jackie, Lucy and me, our incredible amount of luggage and gear, plus a dozen or so two- and five-litre flagons of fresh water in case the creek needed cleaning! Therefore, the dinghy became a displacement craft with a speed of about seven knots, making the journey a tedious thirty minutes.

The tide was very low. Reefs and rocks showed their heads everywhere. So thick were they at one point, I thought at first that we wouldn't get through. The western shore of Tuin – or, at least, the only part of it we could see – was less beautiful than the other side. As the dinghy eased its way through rapidly shallowing water towards the squat hut standing alone on the shore, I was presented with a picture of positive parchedness.

To our left, the shoreline was a very sombre picture indeed, even though shiny in the glare of the unmerciful sun. Huge black chunks of granite and rock, some jagged and pointed, left little sand to be seen on the beach and continued out into the sea. For the entire length behind them were walls of more black granite slabs. The trees and bushes on the higher ground beyond were uninvitingly bristly with brown trunks and dark green leaves. Already I could see that the carpet of tall, coarse grass was turning brown. In places, black rock ledges took over from the grass to provide private and individual beds for any sun-loving reptile. A herpetologist would have shrieked with anticipatory delight at seeing that part of Tuin!

To our right was a greener picture. Our view of the southern part of the western shore was blocked by a curving bay's furthermost protruding arm, culminating in several large rocks set in the water. All round this bay, which one could easily imagine to be a crocodile's play-ground, was tall, dense mangrove with five palms tossing their heads above and behind. A sixth palm couldn't toss its head because, like C. S. Forester's Tribe, it was decapitated and looked incongruously like a telegraph pole!

The centrepiece, wherein stood the lowly shed, was decidedly like drought-stricken Australian bush country. Even the silver-barked gum trees, which dotted Tuin's waist, looked thirsty! There were one or two large trees which seemed to boast that their roots had found water, and they seemed healthy enough, as did the prickly pandanus which vied and jostled with the gums for residence. The undergrowth looked a mess of browny-green entanglement.

I noticed that for once, my loquacious Lucy was strangely silent!

The tide was out at least a hundred and fifty yards over wet, rock-strewn, sand-coloured sand. The dry stretch of beach in front of the shed was brilliant white. To the left I could see what looked like a ditch, and I correctly assumed it to be the creek.

The dinghy scraped bottom and stopped, and we were obliged to walk through hot shallows for twenty-five yards to the water's edge. Mid-way to the shed was the first dry spot – a large black rock – and we decided to unload all the gear on to that, the two crew members helping us to carry it. As we made the first wade to shore, one of the black boys suddenly leapt high into the air with a cry. A poisonous blue ray had shot out from under his foot. It sped by us at an alarming speed, leaving a narrow, sand-blasted wake behind it.

'Oh!' said the boy. 'Very dangerous! Plenty here!' We pussy-footed on through the water and saw several more sand streaks appear to our left and right. Three more journeys each and all the gear was on the rock. Jackie posed Lucy and me, formally shaking hands with the two boys, then the boys departed with alacrity to rejoin the cargo boat. I think they really meant their parting calls of 'Good luck!'

The three of us picked up pieces of luggage and baggage and staggered on. When we reached the dry, white sand, the heat came up at us like a furnace. Jackie and I collapsed in the shade of one of the creek's two trees. Not so Lucy! She tied a silk scarf round her forehead as a sweatband and was marching off to the rock for more gear. 'My God, she's nuts!' said Jackie, an Australian born and bred. 'Doesn't she know that nobody moves about like that in this heat?'

'I can't tell her anything,' I said. 'She won't take any notice of me at all, so I'll just let her get on with it. By the way, what do you think of this place?'

'You certainly couldn't call it a paradise isle. In fact, I think it's pretty horrible.'

'You wouldn't like to try your hand at this sort of thing then?'

'I've been in the bush, done the camping bit. But, no . . . I like my creature comforts too much. This wouldn't be for me. But I admire you two for doing it.'

'Lucy's quite a lady, don't you think?'

'She's rather a strange lady,' said Jackie, 'but she's got guts – that's for sure. I don't think that you two are at all suited, though.'

'Ah,' I said, sadly, 'that's where you're wrong. We are. Unfortunately, for some unknown reason she has decided not to recognise it.'

I was going to elaborate but Lucy, her whole body positively radiating heat and her face the colour of boiled lobster, dumped two more pieces of gear close to us and glared at me. 'Aren't you going to help?' she gritted. 'I hope you are not going to leave me to do everything on my own.'

I gave a deep, ostentatious groan. 'All right,' I sighed, 'let's get everything here.' Rising, I continued, sarcastically: 'You realise, my dear Lucy, that we *do* have plenty of time. Like a whole year, for example.'

Jackie looked at her wristwatch and jumped to her feet. 'We don't!' she exclaimed. 'The helicopter will be here in twenty minutes! Can we put all the gear in one place on the beach? Then I'd like both of you behind it, looking a bit lost and bewildered . . .'

Right on time, the chopper settled on the beach and swept Jackie away into the blue sky. Lucy and I were alone on an island that was as hot as Hades and with a hostility between us that had engendered itself in Brisbane.

However, part of me was contented. I was, at last, on an uninhabited island no matter what its appearance. I lay back in the shade, smoking a cigarette that Bill, the pilot, had given to me. I knew that Lucy was seeing a lazy old man. She wasn't able to appreciate my two philosophies of life in such situations, one born of my Army days: never stand if you can sit, never sit if you can lie down. My second creed I took from a time-worn joke in which an

158

old bull and a young bull crest a hill to see a herd of delectable young heifers grazing in the meadow below. 'Yippee!' shouts the young bull. 'Let's gallop down and fuck a few!'

'No,' said the old bull in his wisdom, 'let's walk down and fuck the lot!'

Perhaps that metaphor is ill-advised in that it could mislead a reader into assuming what Lucy and I were going to do under the tree! Only the platonic sense is meant – for if there is one thing an experienced person doesn't want it is to be told by a novice what *must* be done – not in the gradual order of priority over the weeks, but everything all at once, *now*!

This Lucy did as she peeled off her dress, bra and knickers, rubbed sun cream all over her body and marched defiantly out into the sun's rays to get a billy can from our heap of gear.

Confronting me with proffered billy can in tacit dictate that I should get water, she said, 'Tea!'

My feelings will be far better understood after I have introduced her properly. For this Girl Friday was something more than the other two had been – she was my wife!

Two

A PERSON HAS no need for deep psychological or biological understanding to appreciate that *homo sapiens* structures and personalities are divided into two types. Without writing a *de Profundis* on the subject and not going into the properties of magnetism and certain other qualities, one can broadly say that a person can see in another person an automatic and instinctive attraction, repulsion or non-emotion.

I found many times, for instance, that in a barrack room of twenty men, one fellow-trooper would become my instant, inseparable friend, one I would be forced to battle with until one of us – or both – was knocked out cold, and the other seventeen would be simply acceptable companions.

Type distinction is, I think, more important when boy meets girl, but it is so very often overwhelmed by intense physical attraction, cerebral acumen, proximity at work, loneliness and a desire to procreate.

Reflecting on the women I had associated with after Rosemary's departure, I knew very well that not one of them – even Carol – had been my type at all, at least not for an *intimate*, on-going partnership. Equally, I had not been theirs.

Although Christmas 1980 was little more than two weeks away when I returned from Chile, I immediately set up negotiations with the Seychelles government for an island. It was pointless trying to find another little-known romantic treasure island like Cocos – for I doubt if such a one exists – so I made a bid for the more commonly known.

I then deviated in two ways from my previous pattern for finding a Girl Friday – I specifically stated 'wife', not

left top The western arm of Wafer Bay from a hill above the guards' camp
right top Roddick in Chatham Bay
bottom Pajara Island, with Manuelita Island in the background

Wafer's lone high waterfall

top The creek in Wafer Bay where Benito Bonito supposedly hid his dubloons
bottom Anne Hughes, making a fishing net from her mosquito net to catch sardines

Roddick with the remains of the crashed Flying Fortress

top Lucy and Gerald Kingsland on board the *Torres Strait Islander,* going to Tuin

bottom The author spearing crab for bait in Crocodile Bay

The author with his newly cleaned shotgun

top Surveying saplings to cut for the first shelter
bottom The author in his office, as the first rough shelter became known

top Ronald Lui, his children and one of their friends
bottom Badu women preparing a green turtle

Girl Friday in the advertisement in *Time Out*'s Travellers section, and I was seeking her without having an island to go to, thereby allowing me the opportunity to re-advertise if no one 'my type' materialised with the first one.

Also, and understandably I think, I was a trifle tired of negotiating with governments on my own and I needed someone to inject in me the somewhat lost exuberance and elation of planning for provisions and equipment, and writing a sensational synopsis so that a newspaper would want to buy a feature by paying for air fares and expenses. There was nothing left of my publisher's advance; in fact, air fares to Chile had been paid by the *Daily Express*. Fortunately, there was a good-sized cheque from them waiting for me in return for a feature they had syndicated abroad.

In my imaginings I saw an attractive, intelligent woman who would become a loving companion and partner in the adventure from its very offset.

It was anything but a picture of credibility that I painted to the twenty women I elected to interview out of fifty-two replies. I had little money, no bank account or cards, no property and no collateral. But I must have had something which gave four women I short-listed sufficient assurance to take off with me to an uninhabited island when I had found one!

Whether it was the time of year I don't know, but more upper-class types of women had replied. But which of the four should I choose? I liked all of them!

Should I take the very beautiful, wealthy forty-eight-year-old, the full-bosomed and full-humoured thirty-four-year-old, the demure twenty-three-year-old whose doe-eyes promised passionate tropical nights, or the no-nonsense twenty-four-year-old?

I met each of them twice, never mentioned sex, made notes and sat down to decide. Process of elimination showed a predilection for the twenty-four-year-old Inland Revenue clerk Lucy Irvine. Her voice was well-modulated like Rosemary's and I recognised in Lucy the partner I had been so desperately without since Rosemary's leaving.

But I needed to see all four women again to be absolutely sure. I left Lucy until last.

I knew when I saw her again that she was the one, and I told her so. I felt so comfortably warm with her, her vocabulary matched mine – even contained words I'd let fall through disuse – her wit was sparkling and, through her delightful chatter, I saw a bubbling buccaneering spirit. Also I liked the way she wore her clothes like a model, her ridiculously delicate wrists and her unwavering eyes. I knew completely that she was my type, my kind of girl and I found myself wondering if, under the expensive boots and ankle-length skirt, her calves and ankles conformed to the overall pleasing appearance.

That night I discovered that they did. Her legs were long and shapely like a dancer's. We were at a party of a dear and close friend of mine, Jane Lott, Rosemary's sister, to whom I introduced Lucy as 'my next Girl Friday'. To my surprise, and bringing an arch look from Jane, Lucy got into bed with me when the party ended. It was the early hours of February 1st, Lucy's twenty-fifth birthday. As I commented to her: 'I don't think I could afford to give you any other present!'

Every weekend after that we were intimate, and there was certainly something more than a sexual relationship growing between us. She took me to see her mother, Pam, who lived alone in Teddington, telling her of the island project and announcing that she and I were 'very closely attached'. On March 8th, my fifty-first birthday, Lucy treated me to Vivaldi's *Four Seasons* at the Royal Festival Hall and presented me with a pewter tankard, on which my initials GWK were engraved. It was then I knew I was falling head over heels in love with her.

Lucy certainly helped with negotiations, taking individual days off work to do so. Unfortunately the Seychelles government seemed slow in producing results, and my publishers suggested that I tried an English-speaking government for a change.

Australia's Queensland House, in the shape of silver-haired, Acting First Secretary Allan Gynther, quickly

162

dispelled my growing anxiety: 'We'll get you at an island, mate. No problem'; and he got out the whisky bottle. Telex messages flashed between London and Brisbane. National Parks confirmed there would be an island waiting for us on our arrival. Then Immigration Officer Graham Gillespie, at Canberra House, dropped a minor bombshell: 'You, Gerald, have all the credentials to get you past the three-month immigration queue – I have a letter from your publishers and a confirmed telex from the Queensland government. But Lucy is no relation of yours, you can't claim her as your *de facto* wife – you haven't been together long enough – and she is not even a part of your contract. I can't guarantee that her application will go through with yours. If you two were married, that would be different, and I think my government would feel better if it were party to putting a married couple on one of our uninhabited islands.'

Lucy and I were married at Peckham Register Office on April 3rd. 'I am not *in* love with you,' she had told me, 'but I feel very closely attached to you, and who knows what the year will bring?'

Pam's comment when we told her was: 'Now perhaps Lucy will give me the grandchild I have always wanted.' Her wedding present was forty pounds, most of which went on medical supplies, tampons and spermicide for Lucy's newly-fitted IUD.

The *Sunday Telegraph* signed us up for features and gave me a cheque, and Lucy sold her piano to supplement it. Allan Gynther made a special trip to Heathrow to see us off on April 26th, and he wished us a 'happy twelve months' honeymoon'.

Sydney was ablaze with sunshine. Not that we saw much of it – we spent twenty hours in a hotel bed, getting over jet lag. At least that was our excuse.

In Brisbane, Dr Hugh Lavery, of National Parks, said he had the ideal island for us, part of the Great Barrier Reef and some two hundred and fifty miles from the nearest civilisation, an Aboriginal Reserve. Lucy and I spent many days in the Parks Library, studying flora and fauna. Tim Baker, of my publishers' Brisbane office, wined and dined us at one of the city's leading restaurants, saying we could have the pick of any books, free of charge, to take with us. Lucy was more loving than ever and I was blissfully happy. Then came the blow.

The directors of the National Parks suddenly refused to accept responsibility for putting us on an island and cancelled their offer. Considering our financial position and where we were – on the other side of the globe without return air fares – the news was devastating. Thankfully, Lucy and I were being looked after by Allan's son, Ross, and daughter-in-law, Chris. Also, I remembered another contact that Allan had given me and to whom I had spoken twice on the phone from London.

Frank Moore, chairman of the Queensland Government Tourist Bureau, quickly set the ball in motion again. Within a few days, through his friend Wally Baker of the Lands Commission, Lucy and I were on our way to Thursday Island, air fares paid by the Bureau and our tickets clearly marked in capitals ONE WAY ONLY.

I couldn't help feeling that Australia was trying to get rid of us. We had landed, initially, in the far south at Melbourne, where we and the other passengers were sprayed before going on to Sydney. On up to Brisbane and now we were exiting with a one-way-ticket over the desolate, untamed York Peninsula.

It was on the forever-unforgettable journey, first by jet then by a twin-prop, twenty-seater Fokker Friendship, that Lucy certainly 'fokked' our friendship. More, with her one terse statement at Townsville airport, after we had slept in my small, two-man tent for the first time, she emasculated me and shattered all my dreams.

'I am not having any more sex with you,' she said.

There had been no warning, no justifiable build-up or cause. At first I thought she must be joking. But she was adamant. As I demanded 'Why?' I felt sure I was dreaming. She refused to give a reason and went over to talk to the ticket officer as though nothing untoward had happened. I fetched her back, quickly. 'Lucy!' I said, pressing my hand against my forehead. 'Can't you see what our relationship will be on that island? It will be intolerable. Unbearable.'

'That's your affair,' she said. 'You should have chosen one of the other three women. I didn't ask you to pick me.'

This just wasn't happening. It was too ridiculous. There was no sense, no rhyme, no reason. 'What the hell have I done?' I cried.

'Nothing,' she said, and so far as she was concerned, no matter what I said or did, the subject was irrevocably closed. I felt sick, hurt, rejected and bewildered as we continued our way northwards. For me, the project had suddenly lost most of its meaning, yet I knew I had to go on, for there was nothing else.

In all the time we were together, she would never give me an explanation. All she ever said was: 'You don't understand.' However, I did, eventually, piece together some ideas about her behaviour – but not until many months later.

Love, of course, cannot be turned on and off like a tap, and it would be foolish of me to say that I didn't still love her very dearly. No matter what, I enjoyed her closeness and company, and I rather liked the way she always referred to me to others in company as 'my husband'.

Tropical Thursday Island had its very own archetypal beer-swilling priest, Father MacSweeney. An advance phone call, instigated by my publishers, had him waiting on the heat-cracked wooden wharf as the airport launch drew alongside.

There is always something in the faces of God-fearing men that distinguishes them and, although Father MacSweeney was not in ecclesiastical cloth but preferred a sweat-soaked vest, tatty trousers and slouch-brimmed hat – plus the fact that he had red-tinged eyeballs – I picked him out from the crowd of blacks and whites.

Almost his first words were: 'Let's get out of this heat and have a cold beer!' We threw our heap of gear into his pick-up, the gearbox of which was partly ornamental, and we ground our way to a nearby spreading edifice of patched-up boards on stilts in a moat of broken beer bottles which bore the boastful emblem: *Grand Hotel*. It seemed inevitable that it should be run by a woman called Kate, who was immediately sympathetic to our cause. Trimly decked in sweater and shorts, she halved the price of the double room. Even so, twenty dollars a day was well beyond our means.

'I'll have a woman-to-woman talk with her,' said Lucy purposefully, and she returned to our room all-sparkling to announce there would be no charge after the first night. 'That's absolutely marvellous, Lucy,' I enthused, and felt really proud of her.

To my utter astonishment, she asked me, rather seductively, if I would like to make love to her. A mixture of bewilderment, surprise and pleasure caused me to just stand there. My answer was like a schoolboy's: 'I think I feel too embarrassed to touch you.'

Moving closer, she whispered: 'I really feel like naughtiness' – her pet word for making love. 'Let's undress each other.' And she began pulling my shirt front out of my trousers.

'You're right,' I croaked as we sank, naked, onto the bed. 'I *don't* understand you.'

Physically, I was adequate to enter her, but her damning, categorical statement in Townsville was predominant and I became so on guard with my emotions that I must have gone through the functions with little more regard than I would have had had she been a prostitute. I felt totally inadequate to impart the love and tenderness I really felt for her. She sensed my reluctance, and I sensed her withdrawal. I wanted to say many things, but couldn't. It was obvious she'd had little pleasure when I finished. She flounced out of the room with a backward parting reiteration: 'Positively, I'll never have sex with you again.'

It was then that I seriously considered kicking her out of my life completely before I was hurt any further. But I couldn't. I needed her. Even if I didn't, what sort of man

166

would I be to abandon a woman with only five dollars to her name in a strange land?

And so it came to pass, as they say in Biblical terms, that Gerald and Lucy did not enter the Garden of Tuin hand in hand. And I will leave the reader to assemble and decipher my thoughts as I lay in the shade of that tree, by a pitiful excuse for a creek, with a final cigarette, the memory of which had to last a long, long time.

Three

ALTHOUGH IT was with something akin to dismay that I first viewed our fresh water supply, I was quite relieved when I made a closer inspection. It came as a fairly fast trickle from the bank where the second tree's roots had thrust through, low down. Obviously, the water, quite cool to the touch, ran from a trapped underground pool, perhaps similar to an artesian well. After forming two small, clear puddles – each lined with algae and with vine creepers shooting across from the surrounding vegetation – it cut a curving two-inch-wide rut to the creek's mouth where it dissipated into the beach.

One of my first tasks would be to clean out those puddles and dam them to make them deeper, thereby checking some of the waste running to sea. Parting the grass, I knelt down, pushed aside a few creepers and put my lips to the water. It was beautifully sweet! Thousands of insects swarmed around me as I carefully dipped in the billy, and I saw that they were concentrated in two further puddles which looked revoltingly slimy and stagnant. These, I decided, I would drain as soon as possible and cover with sand.

The shadows were lengthening rapidly as I sat, sipping my tea and listening to Lucy's diatribe of things we '*must* do'.

'I'll get the tent pitched,' I cut her short in mid-sentence.

'What about all our luggage?' she exclaimed.

'What about it?' I said.

'We can't just leave it on the beach!'

'Why not? Who's going to steal it? A crocodile?'

'Supposing it rains?'

'Bully for that – I wish it would. Look, anything you don't want to get wet, put it beneath other things. In any case, when the sun hits tomorrow morning, everything will be dry in about five minutes. This *is* the tropics, you know. *Tomorrow*, not *now*, I'll clean out the shed and we'll use that as a store. I want to feel fresh and alert before I go poking about in there.'

'I *wish* they had been more forthcoming about the snake situation on this island,' she said.

'So do I. Either they are not really sure or they don't give a monkey's about us.'

Having read in Brisbane that a sea crocodile seldom attacks anything on land immediately – unless the female is guarding her eggs – but that it likes to lie in wait, sometimes for as long as two weeks, to study an intended victim's pattern of daily activity first, I decided to camp on the beach in preference to taking a chance inland with death adders in the long grass and shadows.

Close to the ten-foot-wide creek bed cut by the rainy season's torrents was a flat piece of pure white sand on the opposite side to the shed and above the high-tide mark near to where our gear was heaped. I pitched the tent there and it was almost dark by the time I had spread out a double sheet for our bed and placed inside two single sheets for coverings. Seeing me strip off, Lucy exclaimed: 'You're not going to bed, are you?'

'I certainly am,' I said.

'But it's so early!'

'If you want to stay out here and get eaten by mosquitoes, you're welcome,' I said, and picked up my new machete which I had sharpened on a grinding stone on TI.

'What are you going to do with that?'

'Cut your bloody head off if you don't stop talking,' I told her. 'I'm going to sleep with it! Just in case we get a nasty night-time visitor.'

'In that case,' she said, defiantly, 'I shall take the Puma knife and have that by my side.'

'Please yourself,' I said, 'but just look after it, that's all.'

The 'Sea Hunter' knife, kindly donated by Whitby & Co of Cumberland, was the most aesthetic and efficacious' sheath knife I had ever owned. Its long blade was curved like a *kukri*, with a bludgeoned part that could be used as a hammer, and its back was toothed for sawing through bone. Unfortunately, Lucy snapped the blade within two weeks by using it for something it wasn't intended – trying to lever a mud oyster off a rock! I was very upset, but I gave her an all-steel-handled carving knife which fitted the 'Sea Hunter' sheath, which she wore constantly throughout the day on a leather belt I gave her, and she looked quite the Amazon with that and nothing else.

Within a few minutes she joined me in the tent and I could actually feel her hostility as she did so. Carefully, so that not one part of her would touch me, she laid herself down. I was just dropping off when she started: 'When we know the island,' she said, 'I won't stay here with you on this side. I shall sleep outside under the stars on my own. I just won't be able to put up with this, night after night.'

'I wish to Christ you'd go now,' I said. 'I want to go to sleep. *Shut up!*'

A short dream later, I was awake again. 'Listen!' she whispered. 'It sounds like someone being murdered!' A blood-curdling screech brought me more fully awake and I felt the back of my neck prickle. The noise was like the screams of a woman being brutally tortured.

The howls increased in volume. Then they were coming nearer! Fast! Like incoming shells they screeched above and across us, fading into the distance. Then there was silence.

'Birds,' I said, 'but what a bloody horrible, frightening row!'

'I haven't been to sleep, yet,' said Lucy.

'And you're the brave soul who's going to live on the other side of the island?'

'Not because of that!' she said. 'I haven't been to bed as early as this since I was a child.'

'Pity,' I said. 'It might have made you sweeter tempered.'

I next awoke to a lapping, watery sound right next to the tent – or so it seemed – and once more my hackles rose. My senses came fully to, and I realised that the tide had come in. Now and then it emulated the sound of some large fish swishing its tail in the shallows, or even something very big wading out of the water.

'Can you hear it?' said a meek little voice.

'Of course I bloody well can,' I said. 'Shut up. I want to concentrate on it.'

I lay listening, found it had a rhythm which could only be water and it sent me straight back to sleep. Daylight awoke me. Lucy had gone. I unzipped the mosquito flap and saw her, performing calisthenics on the beach; behind her, where the out-going tide was revealing a sand-spit, a majestic row of Australian black and white pelicans stood watching her. Lucy was completely nude, and I could imagine the pelicans thinking: 'What strange manner of person is this?'

After a breakfast of unsweetened semolina and tea – we had deliberately brought no sugar with us – I took the machete and went to the shed. A sheet of corrugated iron was missing from one side, and I could see that the shed had few contents: a once french-polished dining table with a roll of sacking lying on it against the wall, a wooden bench seat, and three plastic and steel chairs, each of which had its back legs missing. Wrenching open the door against a small drift of sand, I saw flashes of movement everywhere. The shed was full of geckos, harmless lizards whose large eyes are as appealing an an English field mouse's. Searching carefully in the corners and under the table I found three bulbous-bellied redback spiders and flattened them instantly with the machete. I had been told to take no chances with those things.

Close to and behind the shed was a large, thick-leafed tree, under which a low table had been made with four empty petrol drums as legs with a square sheet of board on top. There was a long thicket to the left against which were two rickety but serviceable home-made wooden tables. Two further petrol drums had been holed for fires and were half full of wood ash. On the right was another long thicket running almost to the shed door. It looked a snug, sheltered camping spot, albeit a bit overgrown and neglected. I looked under the bushes and up into the tree, and could see no snakes.

'This is certainly a bonus,' I told Lucy. 'We'll make it our temporary camp. Crossfield said we could use the hut if we wanted to.'

After carrying the heavier pieces of luggage into the shed, I left Lucy to see to the rest and to sort out the food and kitchen equipment. I was eager to see what the beach and fishing had to offer. As I had no sun creams – and I certainly wasn't going to ask Lucy for some – I took off my shirt and tied it around my waist so that, initially, I accustomed only one half of my body to the early morning rays, which were already hot.

'We *must* go exploring today, dah-ling!' Lucy called after me. I said a rude word to the breeze and went along the beach, fishing tackle and machete in hand with my favourite wooden-handled carving knife stuck in my belt.

Black-lip oysters were clustered on many rocks; and there was something else, which instantly reminded me of Cocos – *'la cucaracha'*, an elongated, rock-clinging mollusc with a slightly domed shell. Overall, it looks remarkably like a large, legless bug, hence the Costa Rican descriptive name. The cucaracha's tough, orange flesh is excellent bait and is not so easily removed from the hook by crafty fish as is crab bait.

Prising a few off the rocks, I walked back to camp where the water was less rock-strewn. Sand-hugging shovel-nosed rays sped away from me as I waded through the warm water until it was just above my knees. I threw the cucaracha-baited hook as far as I could out to sea.

No sooner had I done so than a three-foot sand shark darted in between me and the shore. I hastily pulled in the line, coiling it like a lasso as I did so, and kept my eyes on the circling shark. Then I threw the line so that the baited hook hit the water a couple of yards in front of the small black fin, slicing the surface. As I expected, the hungry shark bit immediately, and not through the fine steel leader as I feared it might.

Lucy, camera in hand, came running down the beach,

screeching with excitement as I hauled the shark on to the sand and killed it with my knife. I then took the 'Sea Hunter' and made short work of cutting out a few shark steaks for lunch and dinner. But there was such a lot of wastage, for no flesh will keep overnight without pickling, drying or refrigeration. As I threw the bulky remains out to sea, I resolved not to catch any more 'big stuff'.

To my mind it seemed both ironical and just that the first sea-food Lucy and I ate on Tuin should be shark, which, after the first mouthful, she proclaimed to be 'delicious!' Even the flesh we ate on the second day came from members of the shark family. I made a five-tined spear out of eight-inch nails and a long sapling and pinned two shovel-noses to the bottom. Our fish identification book said they were 'excellent eating', and they certainly were.

Besides natural curiosity, the main reason why I wanted to go to other parts of the island was to see if a better camping place were in the offing. Remembering the beauty of the eastern side, Lucy and I set off that second afternoon across the island's flat, sandy centre, immediately after lunch when the sun was still high. I considered it would be better to go in the early morning's cool on the following day, but she became so insistent that I acceded for the sake of peace and quiet. I would have preferred an afternoon's siesta!

So very impatient was she to get started, and she was always a fast walker, that I told her gruffly: 'Go on – lead the way!' She gave me a most peculiar look and went striding off along the edge of the beach, in order that we would circumnavigate a large mound of rocks and thicker vegetation before heading inland. Here, the grass was more sparse and it was easy to see where one was placing each foot.

I have always had the slow gait of a farmer, so I took my time in order to observe the ground, trees and bushes. As we neared the middle of the island, where we could see light through the trees behind and ahead, the undergrowth became thicker, and I called to Lucy who was several yards ahead. 'I'm the one with the machete, Lucy,' I said. 'If you tread on a snake, how the devil can I kill it if you are miles away?'

173

She saw the sense of that and fell in behind me. Some fifty metres from the shore, the grass became so thick that no ground showed through. It was also liberally interspersed with young, very prickly pandanus plants, and somehow I couldn't imagine any snake's wanting to crawl about among them. However, I parted the grass first with the machete before taking each step.

We jumped down the short bank on to the beach by the side of a large, delicately leafed willow pine. Gentle blue waves were effortlessly breaking white for the full length of the beach, and we both considered that part of Tuin to be representative of anyone's wildest dream of a golden-sand paradise island. Unanimously, we called it 'Long Beach'. There, sparkling and glistening white in the centre of the broadly curving bay was 'Kingsland Rock'! Beyond, some four or five miles away, was Moa Island, adding to the general picture of tropical tranquillity.

We took off our footwear and, with the pure and incessant South-East trades buffeting us, walked barefoot in the hot, soft sand towards the northern end.

I would have given anything for our relationship to have been as loving as it had been in London and Brisbane. Oblivious to any feelings I might have had, she took off all her clothes, spreading her arms and offering her face and breasts to the sun.

Bitter resentment churned and welled inside me that this woman should, without reason, have chosen fit to destroy my dreams. More, that she should have instilled in me a blinding sense of loss so shortly after the agony I had undergone with the loss of Rosemary. The resentment was like a malignancy within me that seemed to grow with the days. Yet I did not want to be without her.

At the far end, where the rocks took over from the sand, a giant area of coral reef extended from the shore, and we could see many oysters and clams.

'Dah-ling! We *must* come here with a bucket and collect some!' she shouted above the wind.

Why, I thought despairingly, does she always give voice to the obvious? And why does almost everything she says appear to me as an order as though she, not I, were the one who had lived on an island before? An immense feeling of loneliness, loss and sadness engulfed me. Memories of the warmth and companionship I had shared for so long with Rosemary crowded in, increasing my wretchedness, and I visualised how she would have said those same words: as a suggestion of something else we would share and do together. I didn't want the comparison – I wanted Lucy, not Rosemary. I wanted Lucy to be warm and loving in those idyllic surroundings. But Lucy seemed as far away as Rosemary. Many times after that I called Lucy 'Rosemary', and that did nothing to ameliorate the situation. The word just popped out. When we had been on the island for several months and I suddenly called out to Lucy, 'Rosemary!', Lucy replied: 'You don't love me. You are still carrying a torch for Rosemary.' I tried to explain that seventeen years with a person was a large chunk out of one's life, that my love for Rosemary had ended and that I loved her, Lucy.

After retracing our footsteps we walked on to the other end where thick mangrove bordered the entire southern arm of the bay. *En route* we passed five of the tallest coconut palms that Tuin boasted and, perhaps due to perspective, they appeared to me as the loftiest I had ever seen. I looked longingly at the thick clusters of green pipas hanging below the wind-bent leaves, but I knew that only an experienced native would be able to climb the slim trunks to the dizzy heights. There were several fallen nuts. I sliced one open for a mid-afternoon snack, and we took two others to eat in camp.

Constantly, I was looking for the sign of a creek. There was only one, with a very dry bed, which terminated in a pool of crystal-clear water sealed from the sea by a high, thick bank of sand. I tasted it and spat. Salt!

With the prospect of strong winds drumming our ears daily, and the absence of fresh water, I not unnaturally cancelled out 'Long Beach' as a place to build a hut.

175

Four

WHEN THE tide was out, it was possible to walk across the mangrove-lined bay next to the camp. If my memory is correct – for I kept no diary on Tuin – it was probably the fifth morning that with fishing tackle in one hand and machete in the other I was making my way to the pattern of rocks off the southern arm where the water still lapped around them. In my path were one or two high chunks of black granite, rising up from the already sun-baked sand in the bay itself. As I rounded the second of these I came face to face with a long snout and sprawling, spread-eagled, scaled body. A nine-foot baby crocodile was having a secluded sun-bathe!

It opened its eyes the moment I saw it and we both gave snorts of alarm. Never let it be said that a crocodile is a slow-moving, waddling creature. This one's body shot up instantly on to four stiff legs to whip round and face the other way with a movement too blurred for me to see; then it was galloping like a horse – and almost at that animal's speed – to the mangrove swamp where I heard its crashings through the elevated roots.

'Jesus Christ!' I gasped, and sat down in anticipation of an ensuing heart attack, for I could feel the thudding against my ribs. I continued on, peering anxiously around the other rocks in case 'mum' or 'dad' should be waiting. In particular, I had my eye on a large white rock, sitting a few yards from the granite point itself. I waded through the shallows in between and perched myself on the other side where the water was too deep to see the bottom.

As I sat, fishing and scanning the sea for any tell-tale surface movement, I wondered what Lucy would do should I be suddenly eaten by a crocodile. There would be no trace and she would never know what had happened to me. At first, she might think that I had deserted her; though where I could have disappeared to I knew not. It would always be a mystery. It was quite an alarming thought, and I became more concerned about her anguished loneliness and the subsequent grief of my sons than I did of my own death.

'You are a bloody idiot, Kingland,' I said, and wondered if the sun were affecting me. Nonetheless, I felt infinite relief and pleasure when I saw her fast-browning nude form following my footsteps across the bay. I stood up and waved. Her eyes caught the movement and she waved back.

Naturally, she began talking long before I could hear what she was saying. I think it would be fair and accurate to draw the following parallel to describe my companion's behaviour during the first three weeks on Tuin's parched insularity. After that period of time, though, things did change.

The analogy is something I have never experienced, only heard about – a holiday camp. Apparently, quite often a voice calls out at an unearthly hour in the morning and continues throughout the day, suggesting what to do for the best, and is still announcing the next day's so-called exciting events when campers are trying to sleep.

The only difference between Lucy and the camp announcements was her prefix of everything with a screeched 'Dah-ling!'. Even when she was a hundred yards away, down the beach or wherever, with the wind ramming the words back down her throat, as soon as she thought of something 'frightfully important' she gave voice to it.

It wasn't until she was standing on the rock on the other side of the channel that I caught her words: 'Dah-ling! I've brought your shirt and sun hat.' And she waved them high. As she carefully waded across, she admonished me for leaving them behind. She took my hand and I helped her on to the rock.

'Any luck?' she asked. I pointed to a small pool of trapped sea water on the rock itself in which I had placed a parrot fish.

'Marvellous, darling!' she said. 'I'll scale and gut it, shall I?'

When I told her about the baby crocodile, she didn't appear at all frightened. 'Do you think there are any more?' she asked.

'Probably a nest in that mangrove,' I replied. 'One good thing though, it was frightened of me. If you remember what those books in Brisbane said, the female leaves her eggs after a week to survive on their own, so with any luck she has gone somewhere else and taken hubby with her.' Changing the subject, I asked: 'What have you been doing with yourself while I've been fishing?'

'Looking at your gardens,' she said. 'I never knew seeds could come up so fast. The sweetcorn is just shooting up. So are the runner beans.'

We saw two sand sharks come nosing around and each time I quickly pulled in the line. The poisonous blue rays that flapped gently past ignored the bait completely.

'I want to learn to fish,' said Lucy, and I told her I would make a line up for her, ready for the following day. 'But I do rather enjoy fishing on my own,' I added.

'I don't have to fish next to you,' she said. 'I can find my own rock.'

After one or two very hard bites I pulled in another parrot fish and decided to have one more cast, using the other parrot fish's guts as bait, even though I was fishing without a leader. There was an almost immediate, tremendous tug and the line went slack. I pulled it in. The hook had gone. There, at the break was the small curl that I knew so well – the mark of a shark's bite, and quite a big one at that.

'That's it with that bugger around,' I said. 'Anyway, I think we have enough for today.'

It was then we saw to our dismay that the tide had come in, filling the fifteen-foot-wide channel to a depth of at least five feet with an appreciable current flowing. Also, there was a good-size black-tip shark swimming through it!

'Don't panic,' I said. 'It's easy to get out of this. I will do a flat dive that will counteract the current and take me straight to the other side before anything can bite me.

178

Then you throw everything over to me and I'll be waiting to grab you when you dive across.'

'But I can't dive!' she cried. 'I've never learned!'

'Oh, shit!' I said. 'Never mind,' I added, quickly, 'I'll keep watch from the other side and when I say "go" get down in the water until your shoulders are under then kick out with all your might against the rock. That will send you skimming across as though you had dived.'

'All right,' she said.

I made the shallow dive, my hands hitting the shore rock almost at the same time as I surfaced, and I quickly hauled myself out. Lucy threw over the fish, fishing tackle, knife and machete and I managed to stop them slithering back into the water.

A blue ray lazed by in the current then, so far as I could see, it was all clear. I went down the rock until I was waist deep in the water, held on with one hand and reached over with the other. Seeing that Lucy was ready I shouted 'Go!' With eyes closed and arms outstretched, she pushed hard against the rock with both feet, making a bow wave with her chest. Agonisingly, she seemed to be stationary for a moment, then one of her hands touched mine and I pulled her to safety.

Admiration for her flooded me; but I said nothing, for how could I tell her she was wonderful when she obviously had so little regard for me and even disliked the feel of her hand in mine?

Luckily, there was still a five- or ten-yard swath of sand between the edge of the mangrove and creeping arc of water, and we quickly went round it.

I resolved that next day I would try to find a path around 'Crocodile Bay', as we called it, behind the mangrove so that I could fish from the rock without having to worry about high tide – a much better time for fishing – for I could always negotiate the channel.

Always eager to explore, Lucy went with me and, except for three or four places where I had to hack through with the machete, there was a natural white-sanded avenue between the mangrove and bush, the latter lined not only by the five tall palms but several other smaller ones not visible from the sea.

It had obviously been the true beach before the mangrove took over. Delighted with our find, we named that part of the island 'Coconut Alley'. Bird calls, some weird, some beautiful, were all around us as we walked, sweating profusely in the alley's trapped heat.

When we were more than halfway along, an incident occurred to show me that Lucy did actually dislike touching me. We were suddenly startled by the crushing and splintering of undergrowth and the sound of galloping feet. I instinctively raised the machete and she clutched at me with fear. Whatever the creature was, we never saw it, and thankfully the noise of its running receded in the distance.

Her fright over, Lucy realised that she was holding me, and dropped her arms immediately. I felt quite sad that such a thing should be our situation, especially as once, not so very long ago, those same two arms had enveloped me, lovingly.

'What do you think it was?' she said, stepping away from me.

'How the fuck do I know?' I replied, angrily. 'Probably that little darling I saw yesterday. Or his brother or sister.' And I strode off ahead.

I had made up a line for her, and she sat on the rock with me to fish, infuriating me with nonsense chatter and her pulling the line in every minute or so to inspect the bait. After I had landed two parrot fish and a small snapper she said, sulkily: 'Why do the fish always go to your hook and not to mine?'

'Because they like me,' I replied, belligerently.

'Can't you show me how to do it?'

'There's nothing to show you,' I said. 'Just sit still, shut up talking and stop fiddling with the line.' She couldn't do any or one of those three things – and she caught nothing, much to my smugness and her annoyance.

Naturally, this did nothing to attenuate her almost tangible resentment of my being necessary on that island. I knew very well that it was only my presence and use of the

180

machete that gave her peace of mind to sleep at nights, especially during those first three weeks. Added exacerbation was her having to spend some twelve hours every night in a small tent with a man who didn't want to talk and to whom she considered herself to be no longer 'closely attached'. We had no light for we had no batteries for our torch, and the mosquitos began their marauding at sundown and went to bed at sun-up. There was little one could do but get inside the tent immediately the sun dropped, zip up the mosquito flaps and stay there until dawn. Often, Lucy repeated her longing to sleep elsewhere and I always replied: 'Why don't you?' But of course, she never did.

I think it must have been on our second walk through Coconut Alley that we noticed the giant mound of earth set back a little way in the bushes. It was much higher than we were and some fifteen to twenty feet in diameter. Also, it was riddled with large holes like the warren of monstrous-sized rabbits! Lucy and I stood together in awe and wonderment. What creature or creatures had taken all that trouble to build it? I went forward to have a closer look. 'Be careful, darling,' she called. Ignoring her, I climbed to the top of the mound and began jumping up and down. 'Anything come out?' I shouted.

'No,' she said, with some alarm in her voice. 'Nothing.'

I had a momentary mental picture of something really nasty springing out of one of those holes at her, then, like St George, I would slay it with one blow of the machete and win back the maiden's love! Oh, yes, my mind ran on, the sun has definitely got a hold on you!

'It's as solid as a rock,' I called. 'Perhaps it's the home of goannas.' I peered down the holes but could see only blackness. For the moment it was a mystery to be put aside with one other – large 'pouches' of intricately woven green leaves hanging from branches high up in some trees and bushes.

I have often thought how ironical it was about the sandflies on Tuin, for while I had received many warnings of other dangers, no one – myself included – had considered those lesser, miniscule insects to present any hazard other than a mild irritation. After all, I had weathered mosquitoes and biting insects in many parts of the globe. Unfortunately, I had an allergy to Tuin's sandflies which, in itself wasn't too bad – but there was the poisonous sea to contend with as well!

My allergy manifested itself on about the fourth night. Tiny blisters appeared all over my legs and feet, creating in me a maniacal desire to claw my skin to pieces, bursting the bubble-like excrescences and ridding myself of the intensely itching clear liquid. Lucy remonstrated me for scratching them, but it was impossible to control myself. Each blister produced the same excruciating tickle that a feather does when drawn lightly down the side of one's nose.

Although I applied an iodine-based lotion to the multiple abrasions my nails had left, I noticed, over the next day or so, that a small, yellow core had formed in many of them. Like mosquitoes in most parts of the world, the sandfly is not active in the daytime. Obviously I had to keep off the beach during the night-time. Although Lucy and I shook the sheets every evening, and swept out the sand with a cloth, either of us – or both – had only to go outside for the inevitable midnight pee, and the sandflies would not only attack but invade the tent. To get the tent off the beach was imperative.

I spent almost an entire day at the end of our first week on Tuin clearing an area beneath two trees which stood together some fifty metres inland and afforded shade. Before re-pitching the tent, I turned it inside out and shook it vigorously.

While I was thus engaged, Lucy washed the sheets in the sea. It was as I was helping her to wring them out at the water's edge that we received our first visitors.

That we were a worry to the Queensland Government was soon clear by the way a surveillance plane swooped down upon us every other day, invariably catching Lucy in the nude. As I commented to her a week or two later: 'If that pilot comes in any lower, he'll crash!' Also, every Tuesday and Thursday, Bill swung his helicopter in close to give us a cheering wave en route to the inhabited islands.

On those days, too, the mail plane came flying past.

Because of the noise of the wind and rippling waves, we didn't hear the helicopter until it suddenly came swirling over the trees. And there was a movie camera aimed at us through the open window! 'We're being photographed!' I exclaimed, and Lucy fled to put some clothes on.

The helicopter settled on the beach and out stepped Bill and an ABC television crew. In return for having paid our air fares, the Queensland Government Tourist Bureau wanted a documentary made about us. The television producer, Kenyon Castle, did the interviewing and directed the filming. When I suggested a film about us was a bit premature, he replied that he wanted to capture the atmosphere and our feelings towards the island in the initial stages.

I had dug up small patches of ground in many places for experimental planting and, although the sweetcorn was already four inches high and tomatoes, kohlrabi, melons, pumpkins, beans, chinese cabbage and lettuce were doing well, it did, probably, all look a bit piecemeal.

'This ground is very arid and sandy,' I explained, 'but yesterday I found a patch with earthworms in and that's where I shall do the main cultivation.'

Kenyon seemed fairly disgusted with the amount of provisions we had brought, though I assured him that with fish and molluscs they would just about last us until the vegetables were ready. Even so, our supplies were rather scant – two kilos of dried beans, a kilo of porridge oats, four kilos of brown rice, two packets of dried fruit, one packet of spaghetti, a packet of semolina, cooking oil, tea, vinegar, salt, black pepper.

'What tools have you?' he asked.

'Woodman's axe, machete and spade,' I replied. 'Plus fishing tackle, of course, and some good knives.'

'So you have no rifle?'

'No. Couldn't afford one.'

'You seem to be very ill-equipped for this sort of venture,' he said. 'I don't think I'd like to be here without a gun of some sort. And certainly not with only that amount of food.'

'Because we were *not* put on an island almost as soon as we arrived in Australia – as the government had promised – the small amount of money which we had allocated for food and equipment had gone by the time we did our shopping,' I said. 'But you seem to be missing the whole point of this venture. This *is* supposed to be a survival project, and I know very well that what we have, no matter how little it may seem, is sufficient for us to see it through. The less we have, the more realistic and better the story.'

'Well, I'm pleased that you both seem very confident about it,' said Kenyon, though he obviously didn't share our optimism.

Being a journalist, I knew very well what angle he was going to take, especially when Lucy blurted out: 'Your damn government forced us to marry!'

After they had departed, about two hours later, my anger burst through. 'Lucy!' I said. 'I am sick and tired of hearing about your being forced to marry me and how it was the biggest mistake of your life. I put my trust and faith in you because I was giving up a hell of a lot more than you were – my sons were deeply hurt, Carol washed her hands of me and I lost the friendship of Rosemary's parents. You are a woman of twenty-five and you knew perfectly well what you were doing. So for Christ's sake grow up and have the strength of your own convictions. Stop pretending to be a hard-done-by little girl. No one's interested – and all you are doing is killing the project.'

'What about you?' she stormed back. 'Degrading yourself by asking for cigarettes.'

Angry words apart, Kenyon Castle had put both our backs up, and we were soon treating him as a common enemy, an intruder on our island – besides which, the sheets still needed wringing and hanging out to dry before nightfall!

Captain James Cook called the Torres Strait the 'most unhealthy place in the world'. Over the next weeks, I had cause wholeheartedly to agree with him. My scratched bites had definitely turned into tropical ulcers. My debilitation was unbelievably rapid.

The hard, yellow core of a tropical ulcer is a living organism which eats the cells as it burrows deep to attach itself to bone, creating a deep, pus-filled well with a scab on top. The pain from one can be almost unbearable – as Lucy found when she cut her ankle on coral. At one stage, I had about thirty on each leg. My ankles swelled alarmingly, causing blinding pain and forcing me to lie on my back with both legs raised to drain off the fluid.

Keeping out of the sea was the only answer, which seemed to be almost impossible. The touch of water, even fresh water, was unbearable, so I gave up washing except for my nether regions every morning after going to the lavatory. More important, how was I going to get to and from the fishing rock which was proving a lucrative spot? Lucy and I did try putting a pole between the two rocks and, sitting astride, wriggling our way across. Alas, at high tide, our legs dangled in the water. Fortunately, there was another rock close to camp which jutted out from the shore, and that afforded us an accessible and successful fishing spot without getting our feet wet.

Although it was often painful and I had to rest many times because of my swollen ankles, Lucy and I continued with our exploration of the island, and discovered that the stretch of shore from Crocodile Bay to the southern point was another beautiful white-sand beach with several coconut palms and deep blue water even at low tide. We called this part 'Palm Beach' but, like the southern shore – which had many golden-sand coves divided by black granite promontories – there was no fresh water to be found. It was the same story everywhere on the island – and we knew that our permanent camp would have to be where we were.

Every day, the island and the sea – especially where we were living – became more friendly and familiar. I was also appreciating that there were no really harmful snakes on Tuin – or, if there were, we saw no evidence of them, not even when we walked through chest-high grass on the south-eastern side or through marshy ground behind Long Beach's stretch of mangrove.

For me, Tuin was an uninteresting island, for it would have fitted, with plenty of room to spare, in Cocos's Chatham Bay and could be walked round in a few hours. Yet always, Tuin presented to me a benign, welcoming air as though it would look after me. It wasn't strictly Tuin's fault that the surrounding sea was a killer and that its terrain had, initially, two people living there who were incompatible.

Every night, although we were on the lee side of the island, the wind buffeted the tent, so I decided to build a temporary shelter and put the tent inside it. I loathed and detested my increasing weakness and constant pain, sometimes culminating in dizziness and forcing me to take a rest from swinging the axe or digging the holes for the uprights. Lucy helped with the latter and always went with me to carry back the smaller saplings that I felled. It was while we were erecting the shelter I realised that we could – no matter what hostility there was between us – work as a team.

Raising my arms above my head to place the cross-members and tie them in position exerted more fluid pressure on my ankles. Every time I collapsed in an angry, oath-spewing heap, she would take over the tying to my directions.

There were two other corrugated iron sheds on Tuin – a solid-looking one on Palm Beach and another, which had collapsed, on the southern shore. Lucy and I carried several iron sheets from the second, the heat and exertion taxing us both to the limit of endurance. Four times in as many days we faced that awful journey, climbing over high rocks and trudging through soft sand. Every hundred metres or so we would sit or lie under a tree to allow the bursting heat in our bodies to dissipate. The last journey was the worst. Lucy could drag only one sheet and I could carry only two on my back. At one stage I thought my ankles would explode.

Strange to say, and we often spoke about it later, we really enjoyed the challenge of those journeys. Perhaps it was because the energy we had to expend didn't allow us to think of our relationship!

That first night, after I had nailed on the last sheet with the blunt side of the axe – using bent, rusty nails I had straightened – we felt our efforts were justly rewarded. It was absolute bliss to sleep without the wind banging our ears. Its absence brought more sounds of the night to us, including that of the 'Oh dear' bird, as we named it, for that was its onomatopoetic call.

More and more I wanted to rest each day to conserve my energy in order that my body could cope with the suppurating sores that seemed to have taken over my legs completely.

Apart from when we ate our meals or were fishing – which Lucy thoroughly enjoyed, especially after her first catch, a baby black-tip shark – we spent little of the day together. Now that we knew that the island was reasonably safe, she would go off for walks on her own or sit with her back against a rock on the beach, writing her diary. There was little doubt but that the gulf between us was widening and I felt too ill to care or to do very much about it. The one thing I enjoyed and was determined about was the garden. That was a main means of survival. Survive I intended to do, and I would do it alone if necessary.

It was during the third week that I knew I really was alone. Lucy went down in my estimation to such an extent that I felt I could have no more regard for her. She confessed one morning that she had eaten a whole packet of dried fruit. Shades of Victor, I thought, but this was much worse. Victor hadn't been my partner. There was nothing I could do. I didn't even feel like shouting at her. I just felt hurt and sad.

I knew that soon something would have to explode inside me. It did, at the end of the third week, after two young men on catamarans paid us a visit.

Five

IT WAS late afternoon when we saw the sails come gliding in and stop at the edge of the incoming sea, some hundred metres away. The two men on board were white, so I picked up the machete as a natural precaution and stood by the shed, waiting. Not so Lucy. She'd donned her bikini and was hurrying across the sand before I could stop her. 'Let them come to us,' I shouted. Either she didn't hear or she ignored me. Calling her by another name, I went after her. By the time I reached them, the three had shaken hands and were in animated conversation.

'This is Peter and this is Derek,' she introduced them. 'They've got some mail for us!' She waved and handed me an official buff-coloured envelope.

The two men, like their two catamarans, were slim and clean-cut enough, and I rightly judged them to be in their twenties. Both were Australian and were sailing, as a publicity stunt for the catamarans' manufacturers, from Sydney to Manila.

'One of our rudders is broken,' said Peter. 'Can we stay overnight to repair it? We can give you some food in exchange.'

'That's all right,' I replied. 'But only one night, though. I'm in no position to give any whites permission to land on these islands.'

'We've spoken to Crossfield Ahmat on Badu,' said Derek, 'and he said it would be all right if you didn't mind.'

The buff envelope bore my name and was addressed C/o The Chairman, Badu Island. This told me that the two men were speaking the truth about their having been to Badu. I opened it and found, of all things, a government census form, which brought chuckles of amusement and incredulity from us all!

While I assisted Lucy with the preparation and cooking of the evening meal – to which the two men had contributed tinned meat, tomatoes and half a cabbage – they towed in their catamarans stage by stage with the incoming tide. When they joined us, they put a jar of honey and a packet of cigarettes on the table.

To say the least, it was a very enjoyable meal, and I am sure that Lucy and I ate much, much more than they did! Honey in our tea was a real luxury. That night, Peter and Derek let their catamarans drift on long anchor ropes away from the shore, mosquitoes and sandflies, and slept on board.

Next morning they gave us antibiotic ointment and powder for our ulcers – making a lot of fuss over the one on Lucy's ankle – and half a jar of vitamin C tablets. Then they started to repair the rudder. By mid-day, after asking my opinion, Lucy had dispensed with her bikini top. 'It's entirely up to you,' I had told her. The two men's eyes constantly wandered to her tanned breasts. Lunch over, she announced that the three of them were going for a walk; did I want to accompany them? I declined, then reclined on the grass bed Lucy and I had made by the side of the table, lay back and closed my eyes.

Absurdly, and entirely out of keeping with what I was smoking, a television advertisement crossed my mind: 'Happiness is a cigar called Hamlet.' I also thought of another saying: 'A woman is only a woman – but a cigar is a damn good smoke!' Contented, I slept fitfully.

When the three returned, I saw by the sun they had been away for a good two hours. Lucy looked extremely hot and agitated. I said nothing. After the dried-fruit incident, she was nothing to me. What she did was her affair.

As we sat, sipping honeyed tea, Derek suddenly said: 'We have asked Lucy to sail with us to Manila. We don't think she is happy here.' I was absolutely speechless. Lucy looked extremely embarrassed. 'Well?' I managed to ask her.

'I can't go with them!' she exclaimed. 'I'm your wife!'

'Bollocks!' I said. 'How many times have you declared that our marriage certificate is only a piece of paper? Go on! Clear off with them!'

Turning to the others, I said: 'I think you two bastards have got a real nerve. I have never heard of such under-handedness. But if you want to take her – you're wel-come.'

'I'm *not* going with them,' she cried, jumping to her feet. 'I'm staying here with you.' She ran off down the beach. The two men started to go after her. 'Leave me alone!' she screeched. 'You've caused enough trouble already!' The two men sat down again, sheepishly.

'What have you three sods been talking about?' I de-manded. 'And what have you been saying to upset her like that?'

'Listen,' said Peter, apologetically, 'we haven't said anything against you, but all she has talked about is how badly you treat her and make her do all the work, and that you refuse to gather and chop firewood. She said she is so frightened of you that she sleeps with a knife by her side.'

It was just too unbelievable. 'If she *did* say any of those things, which I doubt because there is no justification – and I think you two buggers egged her on because you fancy her – let me put you straight about a few things. For a start, I gave her the knife. As a matter of fact, it's the second one I've given to her – she snapped the first one – and I sleep with the machete in case something nasty pays us a visit.'

I told them how close and loving Lucy and I had been in London and Brisbane and how I had been really looking forward to a twelve-month honeymoon on a tropical is-land. 'When she told me, virtually, to get lost in Townsville I was shattered and demoralised. I felt like a bee that had been happily humming with what it thought was honey on its knees, then suddenly found it was cow shit. As for her doing all the work – try stopping her! She was so insistent that she knew all there was to know about firewood and wood fires that I just told her to get on with it. When she accused me of using too much wood for boiling water for tea – I did so, deliberately, to put her back up – I never gathered another piece. The fire, wood, stores and cooking are her domain. Fishing, the garden and the creek are mine. I think that's fair enough. I can't *show* her anything, you see. She won't let me. Though I did have to persist in showing her how to use an axe or she would have been minus a foot!'

'Has she ever given you a reason for her behaviour in Townsville?' Peter asked.

'No,' I said. 'Never. It's a complete mystery. Has she said anything to you?' They both shook their heads. 'There has got to be some reason,' I said. 'Perhaps one day I'll find out what it is.'

There was silence for a while. Then Derek said: 'We think you should give up this venture. You're too old and too ill.'

'Listen, you two,' I said, somewhat heatedly, 'ill I may be, but I'm not too old. I know exactly what I'm doing. Just so long as that creek holds out, I'll survive – and without any interference or advice from you.'

'But Lucy says you have done very little to make the camp comfortable,' Peter interrupted.

'Damn it, man,' I replied, 'we've only been here three weeks. What the devil does she expect? I can't create miracles. Look, if Lucy wants to go with you, that's up to her. I shan't try to influence her either way. But I don't think she will. Anyway, for the moment I just want to rest and get rid of these ulcers; and I'm certainly not going to run around like a blue-arsed fly just to please Lucy.'

They persisted in asking me what precautions I had taken should the creek run dry. 'I've put a guttering of bent corrugated iron sheeting along the side of the shelter for catchment if it ever does rain. I know how to make an Arizona still. Father MacSweeney, on TI, gave me a crowbar in case I need to dig a well. Also, survivalists have proved that fresh water can be obtained by sucking raw fish. Then, of course, there are plenty of green coconuts here – when I've devised a way of getting them down! That's enough about me – tell me about your trip.'

They said that several catamarans had started off from Sydney, but the others had given up. I sensed that Peter and Derek were very wary of where they were headed, especially Derek who constantly referred to bandits armed with machine-guns in the China Sea. 'They shoot everyone on board and take the boat,' he said.

'You'll be all right,' I encouraged them. 'Your craft are probably too small for them to worry about.'

Their concern for their safety – and ours – and their many references over dinner to the terrible dangers on islands in the Torres Strait, suggested to me that they might well chicken out before they reached Manila or the China Sea.

That night in the tent, Lucy started to declare, defensively, that she had not said anything derogatory about me to the two men. They had said she was a fool to stay with me.

'I don't know who to believe,' I said, 'and, Lucy, it doesn't really matter. I'm just fed up with you and how you behave – it's all like a bad dream. So just shut up and let me go to sleep.'

Next morning, I slept in much later than usual and Lucy had left the tent before I awoke. I arose and went down to the sea. Peter and Lucy were standing, semi-nude, in a close embrace on the beach. I went towards them. 'I'll thank you, sir,' I said, 'to unhand my wife.' He could see I wasn't joking, and he and Derek knew it was time to leave even though at dinner there had been hints of their staying a further night.

'Lucy!' I said, sharply, 'are you going with them?' Her answer was to stomp by me to the camp. As I watched the sails making for the horizon, I knew I wasn't fooling myself – I was unutterably happy that Lucy had decided to stay, and that we were alone again.

I was busying myself in the garden, prepared to forget the awful incident completely, when she came storming up to me, blathering something about Peter and Derek and how unfair it was to have to live on Tuin with such a person as I. It was time, I knew, for a showdown. My stream of pent-up invective astonished even me. With a liberal sprinkling of expletives, I renounced her claims of probity and integrity, accused her of welshing on the agreement to be my wife for one year, called her a stupid, ignorant little girl, and the most despicable character I'd ever had the misfortune to meet. 'From now on, you filthy little bitch, I'll do what I want to do. Don't you ever dare try to give me another order. You're lower than a snake's arsehole.'

I can still see her standing there, straight and naked, her face full of incredible horror. Then she began to wail like a banshee, fury twisting her face. Saliva bubbled on her lips and ran down her chin. For moments she was inarticulate. Then she went limp and sobbed: 'No one has ever spoken to me like that before.'

'It was certainly time somebody did,' I said viciously.

Streams of tears ran down her cheeks into her open mouth. She sprayed me as she spat her next words: 'I'll never have sex with you again!' Stamping her foot three times in emphasis, she further declared: 'Never! Never! Never!'

I was stupefied that only repetition could be her repartee. I just looked at her and slowly shook my head. 'Lucy, I am sorry to tell you, but you are absolutely fucking nuts!' It was so ridiculous, I started to laugh.

'You've already told me that twice before,' I said. 'We know we're not having sex together again, so why repeat it? Incidentally, there is one favour I'd ask of you.' Her wet eyes looked at me questioning. 'I'd like you to take our marriage certificate,' I went on, 'and stuff it right up your arse. From now on, it's null and void. You're on your own.'

She dashed into the shed and came running back with the very cheap wedding ring I had placed on her finger in London. Mumbling something about making her finger green, she threw it at me. 'You can get your own fucking meals,' she screeched, and went off to the other side of the island.

When the sun had dropped to the place where we both knew there were only some ninety minutes left before dark, she came back. We didn't speak at first and she began, rather ostentatiously, I thought, to prepare a meal for two people. We were very much on guard with each other, but, gradually, almost by Socratic elenchus, we ascertained that, no matter what, we both wanted to stay on Tuin and see the project through.

'I promise I'll be a wife to you in every way, except to have sex with you,' she said.

'All right,' I replied, 'and I'll chop some of the wood for you.'

'I know it's cruel, not to have sex with you,' she continued, 'and I really am sorry. But I am very mixed up and I can't help the way I am. I know I love you – probably always will – but I can't love you as a wife should love her husband. I don't think I can ever be anyone's wife.'

'Nonsense,' I said, 'you'll make a fantastic wife for someone. You're really quite something, you know. All you have to do is let that warm, loving, inner you – which you allowed me to share and enjoy – step outside that guarded, cold harsh veneer.'

'It will take me some time,' she said, 'especially after the things you called me this afternoon. No one has ever shouted at me before, and never has anyone called me such disgusting, filthy names. I never thought it possible to have such a row with someone, then be able to talk to each other afterwards.'

'Words are just words,' I said. 'They are only sounds for communication, and vulgarity is man-made. A cunt could quite easily have been a cup if our ancestors and lexicographers had so decreed.'

'That's really funny,' she said, and, raising her voice to imitate an austere, aloof lady, ordered: 'I'll have a cunt of tea, please!'

'One other thing, Lucy,' I said, 'no more nonsense about your having to sleep with a knife. You know as well as I do I'd never harm a hair on your head.'

That evening, as a token of my shared promise to make

a go of it together, I went along with her suggestion to light a fire on the shore of Crococile Bay and fish in the firelight.

'I'll do the washing-up while you get the gear ready,' she said, excitedly. 'Bring my line as well, spare hooks and weights. Bring the green bucket, we can put the fish in that – better bring your knife as well as mine – don't forget matches – and the machete, of course – we shan't need the axe, there's plenty of wood there that's easy to break – you'd better wear your boots – put your trousers outside and tie string round the bottoms so that the sandflies . . .'

'Lucy!'

'Yes, darling?'

'You're doing it again!'

'Oh!' she said, smiling and pretending to clamp her hand over her mouth. 'I was, wasn't I? Sorry!'

We sat next to the fire, swatting the many mosquitoes that were impervious to the smoke, and trained and strained our eyes on the eerie, gently lapping and sinister water where our lines had been cast in the arc that the flickering flames threw, illuminating and shadowing the mangrove, from which came strange night whisperings, calls and noises.

It wasn't long before Lucy's tongue began moving: 'Dah-ling, can't you see, I want to share adventures with you?'

'Not so loud!' I hissed.

She lowered her voice: 'I know your legs are bad and that you are in terrible pain for most of the time, but I wish you'd come walking with me and not just sit in camp, brooding about those ulcers.'

'I don't think it *has* really registered with you,' I replied, 'that I am in pain nearly every hour of the day. I don't want to go walking about unnecessarily. Besides, this island is no bigger than a sixpence. So what's adventurous about it?'

'I know that Tuin is not like Cocos and the other island you were on, but I feel that Tuin is just waiting to give us adventure. This is adventurous what we are doing now, instead of going to that blasted tent every time it gets dark.'

'It certainly is,' I replied, drily. 'This fire will probably attract every fucking crocodile in the Strait!'

'Oh!' she exclaimed. 'I never thought of that!'

'You remind me of one of my mates in Korea,' I said. 'He was as thick as pig-shit where danger was concerned. When bullets and shrapnel were flying all round us, never once did the thought enter his osseous head that one of them might have had his name on it.'

'Did one of them?' she asked. 'I mean, did he get killed?'

'No,' I said, smiling as I remembered. 'He was a real John Wayne, that bugger. He left about five months before I did. Quite missed him, too. Never saw him again.'

'There you are,' said Lucy. 'He didn't worry and he came through all right.'

'There is a subtle difference, Lucy, between facing danger and deliberately courting it. Also, not being pre- pared for it is just as foolhardy. The Yanks used to make us laugh. They really were cowboys, sitting on a ridge at night-time in full view of Chinky and smoking cigarettes. We weren't at all surprised when a few of them finished up without eyeballs and parts of their heads. A cigarette ember makes a nice little target.'

'Do you want to put the fire out and go back?'

'No. We may as well sit here now and take our chances. Anyway, no creature likes fires, so I think we are pretty safe.'

'What would you do if a crocodile suddenly appeared?' she asked.

'You mean right now, at this precise moment in time?' She laughed for she knew how we both loathed and ridiculed that nauseating and tautological expression.

'Yes,' she replied, 'at this precise moment in time . . .!'

'Seriously?'

'Seriously.'

'Shit myself!'

'I think I would, too.'

'No you wouldn't,' I said. 'You'd thrust a lighted branch into its mouth while I knocked its eyeballs out with the machete.'

'Yes,' she said, 'I think that's probably the best idea.'

We sat there for more than an hour, speaking, like the famous Walrus, of many things. We never saw a crocodile and we never got a single bite – except from the mosquitoes.

'Well, that was a bloody waste of time,' I said, as I kicked sand over the embers.

'No it wasn't,' she said, 'I really enjoyed it.'

Because of cloud, the night was black as charcoal, and we had to grope and feel our way to camp, not being able to see where and on what we were putting our feet. Also, because of what had occurred that afternoon we had both forgotten to check the tent to ensure nothing had crept or crawled in. We always left it open for airing during the day, closing the mosquito flaps just before eating the evening meal.

As I struck a match to look inside, I said: 'If there's a snake in there, Lucy, you can get it out.' But there were only a few mosquitoes, which we quickly swatted. Our faces and hands were covered with bites, and we had also been bitten through our clothing. My ankles had puffed up so much that, at first, I thought I would have to slit my boots open. Lucy helped me to pull them off and the agony was such that I felt that the ankle joints were being torn apart.

I said nothing about how stupid I thought the night fishing had been. Lucy, on her part, never suggested we tried it again. I even held my tongue next day when three or four tiny ulcers appeared on my hands, indicating that it wasn't only mosquitoes that had bitten me.

For three more days she went off to the other side of the island after breakfast and returned in time to make supper. Afterwards, she continued with her daily habit of going off alone for an hour.

During her absences I began to revise and re-type a science-fiction story I had drafted out in Italy.

Every morning and evening, I had to spend tedious time cleaning and dressing my ulcers, enduring each and every agonising pain as cores were scraped out with a scalpel. Sometimes I would feel too exhausted to do anything for about half an hour afterwards. Thankfully, the ones on my hands were killed within a week by the antibiotic ointment and powder which also rid my legs of several.

However, some persisted in an indolent state; one, in particular, was very large and painful.

The regularity and speeds of the Torres Strait tides were always a mystery to me. Sometimes they crawled in, taking hours about it, other times they came in with such speed that I could actually see their advance, almost like a bore sweeping down a river.

One morning I walked across Crocodile Bay to the Palm Beach shed to get some twine I had seen hanging from the rafters. I sat for a while, investigated the inside and contents further – table, crockery and cutlery among other things – cut down the twine, had another sit down in the shade, then started off back to camp. I wore only a shirt and thonged sandals, as it was the turn of my lower portions to be tanned by the sun.

Walking across a particularly slimy patch of coral, I slipped and fell, the thongs on both sandals snapping. Sod's Law took a hand. A sharp-pointed rock dug viciously into the hole of the large ulcer, deepening and widening it, and bringing forth a torrent of blood and oaths. The sun suddenly felt burning hot and I was engulfed to the point of dizziness by intense heat and pain. That wasn't all. I had stayed too long at the shed – the tide was right in to Crocodile Bay! To walk all the way round, through Coconut Alley, was unthinkable. The water mark on rocks had told me that the bay was fairly shallow, so I decided to walk across. The pain as the salt water licked into the ulcers was sudden and blinding, then my legs went numb.

Halfway across, I realised my folly. The water was murky with disturbed sand and almost to my navel. I couldn't see if anything came through the water at me. I lifted up my leg. It was still bleeding. 'This is fucking stupid!' I told myself angrily.

I carried on. A large cloud of sand shot up through the water several feet in front of me and trailed off to the right. Obviously, one of the large brown stingrays had decided to vacate my path. Some of these rays were a yard in diameter, and I wondered what my feelings would be if my bare foot came down on one.

Lucy, having decided to see where I had got to, appeared on the shore. I must surely have presented a most peculiar picture, rising out of the middle of the sea like a miniature Poseidon.

'What are you doing?' she screeched.

'Playing silly buggers,' I yelled. 'What do you think I'm doing?'

'I'll help you,' she said, and marched straight into the water.

'Get back!' I shouted. 'I'm all right – I've got the machete. But I can't see through the water. Get up on that rock then you'll be able to see for me!' One can always see better through water from a high point. I stood still until she was on top of the rock. 'Anything?' I called. 'Nothing,' she shouted back. 'I can see to the bottom all round and there's no movement anywhere. Just watch out for the rays.'

It was funny how really safe I felt with her standing on that rock; I waded on and reached the shore without further incident. When she saw the hole in my leg, she was instantly motherly and sympathetic and stopped berating me for having gone into the sea in the first place.

'You poor old darling,' she said. 'You are in a lot of pain. You sit down with a nice cup of tea and I'll dress your leg for you.'

It was probably then that the thin end of a giant wedge of trust was established between us.

Although we remained courteously distant and on guard

with each other, that wedge grew with the weeks and carried us through the ordeal we were to face. So far as I can remember, apart from a few bickerings and altercations, we had only one more major row all the time we were together. That took place the following week.

Always, of course, I was biased against her, for I could not remove the niggling, nagging resentment within me that all the rows, upsets and unhappiness had been caused by her unexplained estrangement of me in Townsville. Often I brooded; and I mourned the happiness and contentment we could have been sharing on Tuin. I spoke to her of that several times, and even discussed with her the possibility that her sudden changed attitude towards me had allowed the sandfly allergy to invade my body. All she ever replied was: 'I am terribly sorry.'

Six

ALTHOUGH ONE or two sprinklings of rain had fallen – the guttering giving us only a few mugs of tea – each day dawned hot and bright and the ground was soon baking, making it necessary to water the plants night and morning and to build sun shelters for them. The creek, too, had become an even finer trickle.

Lucy started her own small patch of melons and pumpkins which she constantly referred to as 'hers', not 'ours', and which, I told her, were totally unnecessary, especially with the shortage of water.

Things that belonged to me – tent, sheets, towels, rucksack and knives were ours. The camera, which her father had bought her, sun creams, medical supplies and everything in her suitcase were decidedly hers.

She had also brought with her a flute in its own leather and lined case. 'I shall learn to play the flute on the island,' she had declared in London.

If there is one thing that I have always found necessary to give a person learning to play a musical instrument it is plenty of room! I quite expected Lucy to have gone to the other side of the island and had a good blow with the Trades, but oh, no.

Morning and evening I was subjected to the hesitant, tortuous and torturous renderings of Humpty Dumpty, Twinkle Twinkle Little Star, God Save the Queen and such.

Although many times I was sorely tempted to tell her precisely what to do with her instrument, whenever she asked: 'My playing doesn't disturb you, darling, does it?' I always returned a gentlemanly 'No', for I was damned if I'd let her see that she was irritating me.

She did progress remarkably well, for she had an excellent ear for music, and after only four weeks I was able to sing to her rendition of The Wild Colonial Boy. Ironically, a week or so later, when I often enjoyed singing to her playing, she suddenly announced she was giving it up because the mouthpiece had gone rusty. I offered to look at it for her, but she told me to leave it alone.

Once before, when I had moved the flute case to get at something beneath it, she had scolded me severely for touching it.

Even the knife I had given to Lucy was now 'hers' and she had 'her own' fishing line which I had made up for her.

One day I used her line so that I could have two lines out at once. I simply couldn't believe her outburst over my touching and using her property. It was the last straw – especially on top of the sexual frustration that was building up inside me.

'Right!' I exploded. 'If that's the way you want it, that's the way you are going to get it. You've got your stuff – I've got mine. If you don't want to share – to hell with you! From now on, you are not sleeping in *my* tent. You don't want to, in any case, as you have always told me so frequently. You can sleep in the shed with the geckos.

'Can't you see what you are doing to me? Every day I am forced to look at that leering thing of yours – you always sit with your legs wide open – and every night I have to lie next to your naked body without touching it or making love to you.

'Now, it seems, I am not allowed to touch what you claim belongs to you. I've had a gutful, Lucy. Why don't you just piss off to the other side of the island as you keep threatening?'

'Why don't you?' she flung back at me.

'Because, you fucking bitch,' I snarled, 'this island belongs to me. You wouldn't even be here if it weren't for me. I suppose you expect me to clear off and leave you with my sheets, my towel and my garden. Your name is not even on the permission to be here. You are just listed as "one companion".'

202

'You don't want a companion,' she shouted. 'You just want someone to fuck.'

'That was the general idea,' I said, smoothly. 'And I didn't hear you complaining in London and Brisbane.'

We hardly spoke for the rest of the day and I was fully resolved not to go back on what I had said. After the evening meal, she took the flannel night coat someone had given to us on TI for extra sheeting, and marched to the shed. I went to the tent, lay down smack in the centre in luxuriant belligerence, determined to forget her and to sleep. But I couldn't. After some fifteen minutes or so, I arose and walked, naked, through the moonlight to the shed. She had done nothing to make a bed, but was sitting huddled on her suitcase with the night coat around her.

My love just went out to her. 'Come on, Lucy,' I said, gently, 'you can sleep in the tent. I couldn't even treat a dog like this.'

'I'm all right here,' she said, stiffly.

'No, you're not. Don't be so pig-headed. Everyone says things in anger that they don't really mean. If you can't share, I can. Hurry up! I'm being eaten alive!'

She followed me along the short path and I laid on my side of the tent as usual. 'You are weak,' she said, shortly. 'I wouldn't have allowed you back into my tent if I had barred you from it.'

'If you are the epitome of strength,' I retorted, 'then I would rather be weak.'

Slowly, I grew to ignore her feminine attributes, mainly because she did, in fact, behave in many ways like a man, and she seemed to revert to type more quickly than I did. She introduced me to the word 'scatological', both in the excretal and interest in obscenity senses. 'I must go and have a shit,' became her morning heraldic phrase.

She maintained that her vulgarisms had come from me, but I always argued with her on that score. For example, I had never told her to squat and pee like an animal in front of me, whenever and wherever she felt the need. 'You remind me of a little Welsh springer bitch I owned in Italy,' I said. 'She would stop suddenly in mid-stride and

squat for a piss, looking up anxiously at me as she did so.'
Whenever Lucy squatted to pee after that, she would look
up at me and say: 'Woof! Woof!'

Gradually, I lost my civilised embarrassment of de-
faecating with someone performing the same function at
the water's edge a few yards away. I was prompted to say
to her, humorously: 'The family that shits together, stays
together.' Never had I done such a thing or discussed with
a woman the evacuated contents of bowels. Herrick's
poem, *Julia*, in which he declares that his loved-one
'shits!', oft came to mind.

'People who have been married for years and years
never see the private parts and functions of each other as
much as we do,' I said. 'Had I seen Rosemary or Carol
going to the lavatory, I would probably have been re-
volted! Yet I see as much of your tits, arsehole and cunt
as I do of your eyes, nose and mouth and think very little
about it. What an incredible relationship we must have,
Lucy. If other people could suddenly see us or listen to
our conversations, they would be shocked rigid.'

Salvador Dali devotes the entire last chapter of his book,
Diary of a Genius, to the fart and its importance in
removing inhibitions and in creating a more relaxed,
friendly atmosphere. It certainly did all those things for
Lucy and me.

Often, we referred to ourselves as the 'odd couple' and
once I said to her: 'With you, Lucy, it's like having a man
around.' And, dammit, she took it as a compliment.

It was, according to my calculations, during the sixth week
that a fourteen-and-a-half foot aluminium dinghy came
racing towards Palm Beach as Lucy and I walked along it.
Always she carried with her a large piece of orange chiffon
should the sun become too hot on her shoulders and back,
and she draped this around her as a sarong. As I'd had
sufficient sun for that day, the time being around noon, I
was wearing shirt and trousers.

Three black men were in the dinghy which came directly

into shore. The youngest, a slim, good-looking man in his early twenties acted as spokesman. He said his name was Titom and they were from Badu. Pointing towards a group of islands in the misty distance, he added they were on their way there to fish for cray, and would we like to accompany them?

'Thank you,' I said, 'but I can't. My book project calls for me to stay on Tuin.' Then, before I fully realised what I was saying, I added: 'But Lucy can if she wants to.' Naturally, though, I never dreamed that she would even consider it.

To my surprise and extreme perturbation, she said, brightly: 'Are you sure you don't mind, darling?'

'No, of course not,' I replied with a graciousness I in no way felt. Pointedly, I continued: 'I don't suppose it really matters if *you* leave Tuin. It's more important for the book that I stay.'

Seeing my meaning, she said, hastily: 'It's not really cheating if I go. I'm not going to an inhabited island – and I would love to taste crayfish.'

'You need good tucker (food),' said Titom. 'You both plenty skinny.'

'OK,' I said, but I felt very angry and hurt that Lucy could go off with three total strangers and desert me without further thought.

As she stepped into the dinghy, I asked Titom if he had a cigarette to spare. Generously, he gave me a whole packet and a box of matches. As I lay in the shade of a palm, smoking and watching the dinghy speeding away, I wondered if I should have stopped her from going. After all, she was my responsibility. I couldn't, however, restrain an intriguing and amusing thought as I studied the packet of cigarettes. Who had the better bargain? Titom or I?

All afternoon I was rankled over Lucy's desertion, and I became both angry and worried when the sun dropped and no dinghy appeared. Darkness fell. 'Sod the worthless bitch,' I said, and went to bed. There was nothing I could do if either a shark or the three men had gobbled her up.

In my mind, I penned a letter to her mother: 'Dear Pam, I am sorry to inform you, but your daughter has been eaten . . .' Then I fell asleep smiling . . .

The moon was shining when her screeched 'Dah-ling! Dah-ling!' awoke me. Relief and resentment hit me at once. I unzipped the mosquito flap and a plate was thrust inside. On it was a piece of cooked crayfish, about two inches long. 'This is for you,' she said, proudly. 'We've been back about half an hour. They showed me how to cook it. I cut it exactly in half.' She was squatting and peeing when she said the last. 'Shift over! I'm coming in! Your bottom's touching mine!' I wriggled over in a sitting position, munching the cray.

'Isn't it delicious?' she gushed.

'Yes,' I agreed, 'but it must have been the baby of them all!'

'It was only a small portion of the tail,' she pouted. 'It was generous of them to give us that.'

'The way that Titom spoke about our being skinny,' I said, 'I expected you to come back with a load of food.'

'It's their only means of income,' she said. 'You can't expect them to give that to us – anyway, I've had such an exciting time. You'll never guess what I did – I fell out of the dinghy! Can you imagine? I don't know what they must have thought of my clumsiness. But this is the real, exciting part – marvellous material for your book – just listen to this . . .'

'It would appear I have little option,' I said, with a hint of sarcasm.

She told me she was a very privileged white person. She had seen something no other white person had seen – the cave in which were hidden the remains of Badu's two bravest and fiercest warriors. 'They swore me to secrecy – and they allowed me to touch the skulls. I promised never to reveal to anyone where they were and wouldn't go back to take photographs.'

'That really is interesting,' I said, sleepily, and lay down to chew the last mouthful. She prattled on about her 'wonderful day', seeing the men spear the crayfish and their search for turtles' eggs – but I was asleep before she finished.

Two days later, when I was selecting more ground for rotation planting, I found a very large lobster carcase behind a bush. By its condition I could see it hadn't been long out of the sea. I felt very upset by my find, but said nothing.

It was because of Lucy's constantly feeling hungry that she poisoned herself three times on Tuin – the second occasion almost killing her.

She called herself 'Fruit Fly' and was like Eve in the Garden of Eden, oblivious to any serpent that might be lurking as she peered and reached through branches and thickets for a juicy fruit. But one particular yellow plum which grew on a bush in Coconut Alley was as dangerous as any serpent where Lucy was concerned, for she ate the stone as well!

She didn't tell me what she had done until stomach cramps gripped her. I looked in our flora identification book – on no account should the stone be eaten. She spewed and had the runs for almost a day, and screeched at me to go away if I attempted to do anything for her.

Within two weeks, she had poisoned herself for the second time. Walking alone along Palm Beach, her sharp eyes had caught sight of the laden vines of wild runner beans, crawling behind the palms. She came running into camp with her chiffon full of pods. 'Bonanza!' she cried. 'Look what I've found!'

'You haven't eaten any, have you?' I demanded.

'Yes,' she said. 'Twenty-five.' I fetched the book. They were MacKenzie beans – 'highly poisonous if eaten raw – often lethal.'

Silently, I groaned: 'Oh, my God, Lucy, what have you done?'

Stomach pains hit her almost immediately. 'I feel so ill, Gerald, so dreadfully ill,' she cried, and ran to collapse behind a bush where, we later found, a nest of baby pythons lived.

Over the past two weeks, I had noticed more and more

surface veins appearing on Lucy's shrinking body. Where my belly should have been was a wrinkled hole. I had taken a hand with the wood chopping, but I became so weak that it was almost a mechanical action of lifting the axe and letting it fall with its own weight.

I walked round to the other side of the bush and saw Lucy on all fours, yellowy-brown liquid squirting from her, fore and aft.

As I looked at her emaciated body, the fearful thought hit me that this woman was going to die – all because of me and my stupid desire to be Robinson Crusoe. 'Oh, you bloody silly, silly bitch,' I said. 'I love you. I love you.' But she didn't hear me because of the noises she was making. 'Please bury it all for me, darling,' she pleaded, weakly. I did so.

For the five hours before dark the spasms hit her about ten times. I was convinced there could be no poison – in fact, nothing – left inside her. But I was wrong. I awoke in the night, stretched out my hand. She wasn't there! Then I heard her piteous cry: 'Help me, Gerald. Please help me.' I crawled outside and saw her lying in shit and vomit beneath the bright Equatorial stars and an Australian moon that as usual was upside down.

I fetched a bucket of sea water and flannel and washed her. But I couldn't lift her and had to drag her limp body back into the tent.

In the morning when I awoke, she was sleeping peacefully. I went out, quietly, and shook up the bed of grass by the camp table, for the tent would be unbearably hot by mid-morning. All day I made her lie on the grass bed and I waited on her with tea, which she promptly brought up, then rice-gruel and pipa juice which stayed down. Often, she said she couldn't understand why I wanted to help her at all. 'I've been such a bitch to you. I'm so very sorry,' she said.

'Forget that,' I said. 'It doesn't matter. And if you can't understand why I want to look after you, I can't begin to explain.'

The following morning, to my joy, she was decidedly

better. But the poisoning had taken its toll. Her bones stuck out, her breasts were wizened walnuts and her entire body surface was like blue netting. When she raised her head from a lying position, wrinkles from her chin to her chest were as pronounced as the lateral bands on a pharaoh's head-dress. When she raised her body, her skin creased all the way down to her *mons veneris*. I knew, too, that my body was the same.

Typically a female, she said: 'I look dreadful, don't I?'

'Apart from resembling a walking medical chart for surface veins,' I said, cheerfully, 'you look absolutely beautiful.' And I truly meant it.

A vein on the back of each of her calves had varicosed, but I didn't draw her attention to them. 'We are not going to give in, are we?' she said, smiling.

'Good God, no,' I replied. 'We certainly aren't. Those arseholes on TI are not going to win their bets that we don't last three months.' Peter and Derek had told us about the wagers. As I had pointed out, there couldn't be such wagers unless *someone* had faith in our staying the course.

Her high spirits instantly returned and, after discussing the seriousness of our situation, we once more affirmed: 'We'll never give in.' Taking the famous words of Sir Richard Grenville, I added: 'We have not yet begun to fight!'

That our situation was serious, I had no doubt. What always struck me as being incredible – and it still does – was how fast the flesh had gone from our bodies in such a short time. Perhaps Tuin was the answer to all obesity problems! 'Come to Tuin Unhealth Farm,' I would quip. 'We guarantee positive weight loss that will stagger you – and make you stagger!'

Still no rain had fallen – of any consequence, that is – and the July tides had suddenly become higher, salinating the creek. Every morning I had to dig a channel through the dam of sand the tide had heaped to drain away the salt water. Then I would sit for an hour or more, gently dipping a mug, tasting the water and pouring it into flagons for storage and putting sufficient in a bucket for that day's needs. On one occasion, as I was sitting there, waiting for the puddle to run fresh, I saw a movement in the grass close by, and found myself looking into a pair of very shy and hooded eyes. They belonged to a non-poisonous six-foot goanna, its body sinister black and flecked orange and yellow. Immediately its eyes caught mine, it whirled around and fled. Several times after that, we saw goannas in and near the creek. Obviously, they were becoming as thirsty as we were.

We did have another, minor, source of liquid for I had devised a way of getting the pipas; three long saplings lashed together with a fork at one end, through which I hung the centre of our very long rope, the idea being to loop it over the nut and pull it down. Bathed in sweat from scalp to feet, and with veins bulging along my arms and across my chest, I made the first attempt. But the weight of the rope made the pole bend like a giant fishing rod and it took every ounce of remaining strength within me to flick it up, balance it and get the loop over the nut. Just before I managed to secure the rope in position on the tendril Lucy pulled the ends, bringing the pole, not the nut, swinging and crashing down to miss her shoulder by inches.

I hurled a stream of expletives at her, causing her eyes to open wide in hurt amazement, then I collapsed with laughter at the ridiculousness of the situation. My laughter eased the tension in my shrivelled muscles like a Salvador Dali fart, became infectious and Lucy collapsed with mirthful relief as well.

After that, not only were we successful in bringing the nuts to earth, but Lucy learned to ignore, and return, any foul names I called her. More and more we were working as a team.

Although I had tried to keep to my vow that I wouldn't catch any more 'big stuff' when I fished, with the shark nuisance one couldn't avoid getting something big on the hook. During the first week I had caught another shark, about four feet in length, when I was using a leader.

Landing it on the beach, I waited until it was dozy through lack of oxygen, wriggled out the hook – which was only through the side of its mouth – and dragged it by the tail back into the sea. Then I pushed it through the water until it revived, and it swam lazily away.

After that I always fished without a leader so that any shark which took the bait, bit off the hook as well. At first, I hadn't much minded this, but now our fishing tackle was very much in short supply with all the swallowed hooks, bitten-off line and hooks caught, irretrievably, on the coral. So when a five-foot sand shark persisted in robbing us, I decided to get it.

Lucy and I were now fishing regularly from the rock near the camp which we called 'Armchair Rock' because of the 'comfortable' shape of a portion of it. I took a brand-new blue nylon washing line with a metal coat hanger for a leader and attached the one remaining large hook I had brought from Robinson Crusoe Island. 'The bastard won't get away from this,' I told Lucy.

Three days later Lucy had left me on the rock to put the tea billy on when the shark came round again. I quickly stuck a small parrot fish on the large hook and threw it out. The bite was almost immediate. I had tied the end of the washing line round a column of rock and it held with the tremendous jerk. The shark shot out of the water, arching its body, and crashed back with such force that two black sucker fish were torn from its body.

Taking hold of the taut line, I braced my bare feet and began pulling. The shark never gave an inch.

Looking down at the wrinkled skin on my puny arms, I mourned my lost muscles. The shark was heavier than I was. By sitting down, wedging my feet against a higher piece of rock and leaning my pitiful weight backwards, I was able to bring the shark in a few feet at a time. Then my little-remaining strength suddenly left me. The line was slowly being pulled back through my hands.

'Lucy!' I almost screamed. 'Help me!'

She sped along the beach and together we hauled that shark in, our bony bodies jostling together. When we had the shark's head resting on the rock, I looped the line round the column of rock and told Lucy to keep it taut. As I picked up the wooden-handled carving knife which I had sharpened to a needle point and walked unsteadily towards the rows of menacing teeth, she cried: 'Oh, be careful, darling!'

I waited until the shark stopped swinging its body for a moment then plunged the knife in. To my surprise it went in quite a way, and I hammered it in further with the heel of my hand. It was necessary to decapitate and gut the shark, and carve its body up into six sections before we could carry it, a piece at a time, back to camp. Sinking two sapling trunks as posts into the beach some fifteen feet apart and above the high water mark, I strung a line between. All afternoon, I cut the flesh and meat into strips. Lucy threaded each piece with twine and hung it on the line. We retained two large chunks for supper.

Three days later, with just the sun and wind, the strips were dried. For nearly three weeks, Lucy made various tasty meals with them.

On one of her lone walks to the southern part of the island – or it could have been on the afternoon she went there with Peter and Derek – Lucy had seen what looked like two gardens but couldn't remember their exact locations. I set off with her and, after an hour or so of searching, she spotted one amid tall grass when she had climbed to a high point, leaving me to lie down with my aching ankles raised. It was obviously a cultivated garden of sweet potatoes! A veritable Godsend. 'I'm sure that whoever planted them wouldn't mind if we took a few,' I said. On our hands and knees we uncovered a few of the mounds, the hot, baked soil biting deep beneath our fingernails, removed the delicious-looking tubers, taking only one or two from each plant, and scraped the soil back over the remainder so that no one would know we had been there. Lucy found she could cook them in sea water, and it was heaven to be eating a different form of carbo-hydrate. Sometimes we would eat them with nothing else for the mid-day meal.

Every three days or so, we made our almost guilt-ridden pilgrimage to the garden and, like the original pair, took the forbidden fruit. Whenever I felt too ill, Lucy made the journey alone. Sometimes, at her insistence, we would make picnic excursions of the expeditions, taking with us a billy can and matches. I would build a fire and boil salt water in the billy while she got the potatoes. Then, with our bellies full, we would return with sufficient to eke out our dwindling stores for a day or two.

Invariably I would exonerate our stealing with the mitigating statement: 'When we leave Tuin, I will donate to the gardener, whoever he is, a lucrative and beautiful garden in complete repayment and compensation.'

Although we estimated our calorie intake to be under a thousand a day, I always told Lucy she was a genius for the way she produced variety with rice, porridge, hard beans, mungo beans, coconut, sweet potatoes, fish, fish, fish and fish. There was no doubt about it, though, we were both becoming a little nauseous over the sight and taste of fish.

One day, I declared: 'I'm going to get a goanna. We need meat.' She accompanied me out into the bush behind Coconut Alley where we often saw the black variety. Rounding a thicket, I almost fell over a sleeping one. Its body looked hideously and glisteningly poisonous in the strong sunlight. As I raised the machete, the creature's eyes opened – and it was ten yards away before the blade descended to dig itself in the ground.

'It's no good, Lucy,' I said, despairingly, 'I'm not fast enough.'

We had many small lizards as camp pets, two of which were very tame and loved to lie full-stretch on the uppers of our bare feet. One had a fat belly with an orange chest, and we named it 'Bronzy'. The other, long and silvery in colour, we called 'Oscar' (Wilde). We contemplated them in our hunger, but unanimously decided 'No!'

While our bodies were weakening, the closeness between us was strengthening. Seeing one morning that I was worried about the vegetable plants, which appeared to have stopped growing – and quite a few had succumbed to the heat – she cupped my face in her hands and said: 'I love you. Don't you ever forget it. We are going to make it.'

'If only it would rain, Lucy,' I said. 'If only it would rain.'

When we crawled into the tent each night, the skin on our bellies hung and swung like an old bitch's dried-up udders. We played the word games I had played with my sons on Cocos, and sometimes we would lie holding hands. 'Geriatrics making love,' she quipped. Seriously, she added: 'I really am sorry about the sexual side.'

'That's all right, Lucy,' I replied. 'I couldn't do it now if I tried.'

There was another game in the tent at night: fantasising to each other about being invited to a thirteen-course banquet, enumerating each and every lascivious dish, invariably starting off with a whole loaf of bread and a pound of butter each while waiting for the antipasto or hors-d'oeuvre to be served! It was fantastic fun, too, choosing and ordering the wine for each course, mine in Italian, hers in French.

Another culinary caprice was the dropping by parachute of a bumper hamper, supplied by Fortnum and Mason, from the passing mail plane. We actually argued once over whether we would drink both bottles of port after the feast, or save one for the next day! Although, ordinarily I never eat confectionery, I suddenly had a craving for fruit and nut chocolate. Always, for my part, the hamper contained a pound bar. 'I think I must be pregnant,' I said.

It was then that Lucy confided that she was almost sure she was. She hadn't had a period since leaving Brisbane. I related the story of Anne, adding that Anne's periods had re-started almost as soon as she was back in England. 'Her doctor told her it was not an uncommon thing with change of food, circumstances and climate,' I added. But Lucy insisted on knowing my intentions if a baby were on the way, for she would never have an abortion.

'It seems to me a bit of a cheek to ask me that,' I said, 'when we are not actually living as man and wife.' But, without any hesitation, I knew what I would do. 'Stand by you and give it a name,' I said, truthfully. We even discussed the possibility of staying on Tuin – or, at least, in the Torres Strait – and raising the child until it was of school age.

Not only was the ground becoming even hotter but the meagre amount of fruit that Tuin afforded was decidedly dying off. One of Lucy's delights, the ubiquitous small, wild passion fruit which we both ate daily, had come to the end of its season's run. The camp tree, under which the table stood, was a plum with a peculiarity for dropping only one of its cleverly-hidden fruits each day, earning from us the name 'One-plum-a-day-tree'. This did continue with its ephemeral donation to flavour, slightly, our breakfast rice.

More and more I was receiving a palate-shocking mouthful of salt as I tested the water in the creek. Sometimes, three days would elapse before it regained its freshness, forcing us to drink from our bottled storage. At those times, too, I would spare water for the plants in the early morning only.

Never did Lucy touch the machete, which she considered to be a highly dangerous weapon, and always she would sit, longingly, her tongue trying to moisten parched lips, as I took many, many chops to cut the end of a pipa. No more could I hold the green nut in my left hand and bring the blade deftly down about an inch or two from my thumb, neatly slicing through the husk with only three cuts. Although my aim was still good, the force behind the blade was pathetic.

'I think I expend more energy and moisture getting the nuts down and cutting them open,' I complained, 'than the juice puts back into me.'

'I *wish* I could do it,' said Lucy, 'but I'd just take my thumb off.'

Seven

NEVER COULD I recall having seen such a footprint. The deep, forcibly made mark was in three distinctive parts, like the composition of a giant insect. The indentation of the heel described the body, the ball shaped the thorax, and the big toe, jutting a full three inches, formed the elongated head.

Had a native from one of the neighbouring islands paid us a nocturnal visit, his presence betrayed by that Gargantuan imprint in virgin beach before the early-morning incoming tide could smooth it over?

No. For the momentary shock of finding such a footprint on the beach outside the camp was almost instantly cancelled by a humorous realisation, depicted in my mind by a parody of the old music hall joke – 'That's no native – that's my wife!'

It was Lucy's left foot that had descended heavily on a more accommodating piece of the hard, sea-settled sand in her purposeful dawn march to the water's edge, and the first time she had deigned to discard her leather sandals on the beach. Naturally, the giving sand had exaggerated the size of the print. Even so, the grossness of it – especially that of the big toe – was, initially, quite alarming.

Deciding to tease Lucy, I dashed back to camp, exclaiming: 'Come quickly! A yeti, or something like that, has been on the island!'

She took the joke in good part, though she, too was surprised at the size of the mark her foot had made.

'If you walked bare-footed as I do,' I suggested, 'perhaps those large bunions of yours would go. They're obviously the result of wearing tight shoes when you were young.'

'I don't think that will help much at this late stage,' she said. Nonetheless, after that she discarded her sandals more and more. As our fish identification book informed us, stonefish lurked behind a receding tide in muddy pools. All we had to do to avoid them was to walk on dry sand.

I was slightly disappointed that I had not found the famous footprint that had left Crusoe 'thunderstruck'. But a month later, in early August, I did find, unmistakably, strange footprints in our beach sand. Before that, however, I was to witness another incident, involving Lucy's feet. It occurred when we were walking along the southern beach where coral also joined the island. Striding ahead of me, as she invariably did on the beaches, she had not seen a baby turtle – a retarded hatching – that, like its previously-hatched brothers and sisters, was making its death-defying waddle to the sea.

The ball of her foot had embedded the poor little thing in the sand, its tiny flippers and head outstretched like a dead goose all around it. I picked it up, dusted it off and it came to life. 'I bet you wondered what the devil that was, descending on you like a ton of bricks,' I cooed to it. Holding it up, I called to Lucy: 'Look what you nearly killed!' She came running back and made a fuss of the 'sweet little thing'. Feeling rather like Androcles, I placed the baby turtle in the water. It dived instantly, came to the surface a few yards further on, took a gulp of breath through its tiny mouth and dived again. We watched it doing that until it was out of sight. 'I hope that one survives,' I said. 'It certainly deserves to after having your foot flatten it.'

It was when we had returned from gathering oysters on the eastern side that I discovered the strange footprints, leading from the sea into camp. There were two sets, one those of a child, the other those of an adult with a heavier and larger frame than the slim-bodied Titom. Lucy and I inspected our things and found that nothing was missing or disturbed.

'They'll probably come back again,' I said.

It wasn't until the following week that the small mystery was solved – and my Man Friday, at last, materialised. It was quite a startling figure that suddenly confronted me as I sat beneath 'One-plum-a-day-tree', grating coconut for the next morning's rice while Lucy had gone for her customary short walk.

His eyeballs were the colour of fire, set in a fleshy mahogany face and crowned by short, crinkly grey hair. A huge belly hung over stretched gingham shorts and his black, bulky frame was carried on chunky bowed legs. He cut anything but a dashing figure. He had come upon me silently. As though the shock of his unexpected presence were insufficient, behind him, with their arms folded across their massive chests, stood two muscle-bound, jet-black young men, regarding me with cold eyes. I measured the distance to where the machete leaned against the tree.

'That my shed,' he said, aggressively and without preamble or introduction. 'You steal my tools.'

I assured that I had taken nothing from his shed, adding that the chairman of Badu had given me permission to use it. 'I go look,' he said, and marched off to the shed, leaving the other two watching me. He returned with the roll of sacking and his belligerence had disappeared completely.

'I'm sorry,' he said. 'My tools here. Suitcases on them. I no see them when I come here other day. You not here, so I not touch your stuff in shed.'

'So they were your footprints,' I said. 'Who was with you?' 'My small son. Friends bring me here this time.'

I invited him to sit on the table, and the other two sat down on the sand. He told me his name was Ronald Lui, that he lived on Badu and that every year he grew melons on Tuin. 'You pretty sick man,' he said. 'Where's your womans?' Talk of the devil, I thought, for at that moment Lucy appeared through the gums and pandanus. Luckily, she had seen them first and had put on her chiffon. Recovered from her initial shock of seeing them sitting with me, for they all three looked quite formidable, she quickly said with true-blue hospitality: 'Tea?' They grunted and nodded.

While the two young ones said not a word, Ronald wanted to talk and I noticed that as his shyness left, so his English improved. I told him about our project. 'We've been told to keep away from Tuin,' he said. 'We believed you were government agents sent up here to spy on us.' Lucy and I roared with laughter at that.

Ronald also cleared up another small mystery for me – the large mounds of earth with giant holes in them, for we had seen several others apart from the one in Coconut Alley. 'They are made by the scrubfowl, and in each hole, the hen lays one egg. Very difficult to get for the hole goes in many directions not straight.'

'But the mounds are tremendous,' I said, in amazement.

'Birds work very hard,' he said.

'But it must take them months?' I went on. Ronald just nodded.

I told him about the baby crocodile. 'Probably gone now that you are here,' he said. 'Crocodile not come much on Tuin. Next island, though, called Wia,' and he pointed in the direction of Badu, 'plenty crocodile there in rainy season. They breed there in freshwater swamp.'

'What about snakes on Tuin?' I asked.

'I have never seen poisonous ones, but there are plenty of carpet snake (python) and whipsnake. They no hurt you.'

'Stonefish?'

'A few. But coral dangerous too. Watch where you put your feet.'

'I wish I could have spoken to you before,' I said. 'Then I would never have camped on the beach in the first place. Sandflies caused all these ulcers.'

As Lucy handed out the mugs of tea, Ronald studied her. 'You two plenty sick people,' he said. 'Both too skinny!'

'My husband is very ill,' said Lucy, pointing to my legs. 'Can you take a note to the nurses on Badu? We need medicines.'

'You write note, I take it,' said Ronald, and took a sip of his tea. 'Ugh!' he exclaimed, and threw it away. 'You drinking salt water!'

'What!' I said.

'You no notice? Very bad for you. You both plenty sick.'

It was obvious what had happened. Our palates had

been adjusting themselves to the taste of salt. 'How awful,' said Lucy. 'And we never knew.'

'You try the well on Tuin?' asked Ronald. I told him that Crossfield Ahmat had said there was a well on the island, but he didn't know where.

'Come with me,' said Ronald, rising. 'I show you.'

I followed him to a dense clump of sharp-bladed grass and young, prickly pandanus plants. Sure enough, in the centre there was a crumble-sided hole about two-feet deep and as dry as a bone.

Only fifty metres from camp, it was a place we had always avoided. As I commented to Ronald, it didn't look much like a well in any case. 'You dig down a few feet, you find water,' he said.

Before leaving, Ronald tasted the water in the flagons. Some he rejected and some he confirmed as being fresh. 'I bring you fresh water very soon,' he promised, and took Lucy's note with him.

Immediately the three had left, I cut a wide path to the hole and began to dig it out. But there was little time left before dark. First thing next morning, I continued digging, shoring the sides to stop the dry, sandy soil from falling in. By mid-day, and despite many rests in between, I was down to shoulder height and nearly swooning with the effort. But the ground was moist beneath my feet! Another spit – and water began to trickle in. I shouted the news to Lucy.

Even as we watched, the water gradually turned a creamy colour and did not look at all tempting. We strained it twice through Lucy's chiffon, boiled it and then re-boiled it to make tea. It had a most peculiar taste.

'If Ronald brings us drinking water,' I said, 'we can give this to the plants. I'm sure they won't mind what it tastes like.'

Two days later, Ronald returned – not only with two five-gallon canisters of fresh water but with the two Australian nurses from Badu. Hearing Ronald's description of us, they had decided to visit us. Our mouths watered when we saw oranges, lemons, jar of honey, butter and a couple of tins of meat.

But Janine and Bunty, as they introduced themselves, were clearly horrified when they actually saw us. They wanted to take us off Tuin immediately. 'Gerald, you should be in hospital with those ulcers,' said Janine. 'I have never seen such a mess. Any doctor would have no hesitation in ordering you to bed so that your legs can have proper medical treatment.'

'I am not going to hospital,' I said, 'and that is definite. This project means too much to me for that.'

'What about you, Lucy?' Janine asked.

'I want to stay on the island, too,' she said. 'Surely you can give us some antibiotics and dressings for Gerald's legs?'

Seeing that we were both adamant, Janine relented. She started us both on eight-day courses of antibiotic tablets, iron and vitamin pills, and showed me how to tend and dress my ulcers. 'I will leave you plenty of dressings,' she told me, 'but you must keep out of the sea at all costs. Ronald has promised to keep an eye on you and we'll see how you are after a week. But I would be happier if you were in hospital. I can see by the inside of your eyes and from what you've told me about liking a drink that a severe alcoholic withdrawal fight has been going on inside you, and that lowered your resistance.'

When the two nurses had taken Lucy along the beach for a urine specimen for pregnancy testing and to palpate her stomach, Ronald handed me his tin of rolling tobacco and papers. It was as we sat smoking, and after Ronald had proudly told me that he had served in the First Australian Regiment in World War II, that we touched on a subject that was to change the whole Torres Strait venture for Lucy and me. I told him of my Army training on Rolls-Royce engines. 'I have machine for you to repair,' he said. 'I bring it next time I come with water.'

The two nurses returned with Lucy, virtually convinced that she wasn't pregnant. We chatted together for a while, then it was time for them to leave. 'Good luck,' said Janine and Bunty. To Lucy, Janine said: 'Look after him. Make sure he eats as much fruit as you do, if not more!'

That evening, Lucy prepared and cooked the most delicious meal we had ever eaten on Tuin – tinned meat with buttered sweet potato for the first course, and a mouth-watering dessert of caramel sauce – made with the honey – boiled brown rice and quarters of orange stuck in the top. Many times I had complimented her cooking, and once I had told her: 'You make better meals with a blackened billy and wood fire than most women do with a de luxe cooker.' Now, I was almost speechless over her prowess, and could only breathe 'My God! My God!' between each savoured mouthful.

It was in the tent that night, as we laid side by side with our bellies feeling unaccustomedly full, that I was forced to dampen, slightly, our satiated states by giving voice to a new concern within me. With the advent of Ronald, the nurses, medicants and food, was I looking at another failed attempt to be Robinson Crusoe?

'Nonsense! Of course not!' said Lucy. 'Think of it as being the shipwreck that Crusoe was given.'

'When Janine was dressing my legs,' I went on, 'I asked her if she thought we would have survived. She said that by doing very little work and exercise, and resting most of the time, the ulcers might have gone away, but not for several months. Our drinking salty water was a different matter, though. We might not have realised we were on the way out, and simply died before we could do anything about it. The salt would have addled our brains.'

Lucy didn't answer for a while, and I gave a short laugh with the most nagging thought of all. 'You can't get a much bigger failure than death!' I said.

'The only answer I can think of,' said Lucy, slowly and deliberately, 'is that Crusoe would probably have died, too, without the shipwreck. Defoe obviously couldn't see how Crusoe could have survived without it. Don't worry about it. The main thing is that we are still on Tuin.'

As I had no wish to upset the present cordiality, I held back from mentioning that she had already been off the island for almost a day.

Within a couple of weeks, she went off in Titom's dinghy

again, this time returning at dusk with several giant pipas and some turtle's eggs. The tide was far out, and I helped her to carry them across the sand, making several journeys.

She appeared to be so animated and excited at 'bringing home the bacon' that again I restrained myself from saying anything. Later, I did suggest that it might be better if she didn't go on any more excursions as we no longer needed food. She was instantly angry, saying: 'I like the pipa juice and I'm fed up with having to ask you to fetch them.' But she never went off again.

I couldn't believe it when Ronald brought his 'machine' a few days later. It was an old Singer sewing machine! What I didn't know about sewing machines would have filled a book. Nothing ventured, nothing gained, I thought, and took it all to pieces with the tools Ronald had also brought, just to find out how the damned thing worked. Every part was filthy, but otherwise it was in good condition for its age.

Ronald watched rather sceptically as I cleaned each part and laid it on the table. Then I put it all back together, oiled it from a can Ronald handed to me, re-set the needle so that it didn't jam with the bobbin as it had done previously, threaded the needle and tested it on a piece of cloth Ronald had also brought in anticipation of my claimed skills.

A few adjustments of the tensioner and stitch spacer, and it worked beautifully! 'Your husband damn good mechanic,' said Ronald to an equally delighted Lucy. 'He's a man we very much need here.'

And that was how my role suddenly changed from Robinson Crusoe, survivalist, to Robinson Crusoe, mechanic, with his own brand of Man Friday who brought him food and water in exchange for repairing Islanders' engines.

The antibiotics worked very fast. If my debilitation had been rapid, my return to health was, it seemed, like lightning. Every day, I felt my strength returning and soon, blissfully, I was without pain and swollen ankles. With the number of engines I had to repair – each one a decided challenge through neglect and rust – I almost forgot Lucy's presence.

It was wonderful to be healthy and be able to think straight again; and foremost in my now clear way of thinking was a deeper resentment for Lucy's cavalier treatment of me over the past three months. I was well and strong again, living a life I thoroughly enjoyed, and each engine that I brought back to life gave my life a new meaning of achievement and philanthropy. Small wonder that in my mind, Lucy was fast becoming incidental.

She and I had previously calculated that we were far into September and were very surprised when Ronald told us it was only mid-August. That quite pleased me, for I had, for some unknown reason, always cherished the idea of celebrating our hundredth day on Tuin, dressing for the occasion and possibly sipping a glass of wine, though where the clothes and wine were coming from, God only knew! I told Ronald about my fantasy and asked if he could bring a bottle of wine as well as water. 'I have a bad stomach and I don't drink,' he said, 'but I will see what I can do. When is it?'

'If we landed here, as I think we did, on May 22nd, the hundredth day should be August 30th,' I said.

'That's only five days away. OK. Leave it with me.'

Just after lunch on August 30th, it wasn't Ronald who arrived with wine but Titom and two other Islanders, one of them a giant of a man. Placing two flagons of white wine on the table he said: 'This for your hundred-day celebration.' But he wasn't leaving them for Lucy and me to drink that evening. Oh, no! *I* had to drink it, there and then, with them! Ronald had told me that Islander women who drank were not very nice, and Lucy was not invited to drink. I explained that Lucy was white, that we had different ways about that sort of thing, and she was allowed to join in.

The wine soon hit all of us, and the Islanders began falling about, Titom casting sly looks at Lucy. Suddenly, he challenged me to arm wrestling. It was easy to see he wanted to impress Lucy. Shades of Victor yet again, I thought, the difference being that here was a young man who led a vigorous life, and he fancied my wife.

Naturally, I had to accept the challenge. For many minutes, our arms were locked, upright, muscles straining. But I could feel his stamina going and I knew I had him. Moments later, I sent his fist crashing to the table. I did the same with his other arm. 'You pretty strong for old bloke,' said Titom. Like Victor he bore no resentment, the incident was instantly forgotten and we carried on drinking until the wine had gone. Then they departed.

That night I couldn't stop singing for about three hours. Lucy interrupted me once to say: 'I was really proud of you this afternoon. Every fibre of my being was willing you to beat Titom.'

I chuckled. 'Bloody good job that big chap didn't challenge me, though!'

A few mornings later, Titom visited us alone to tell me that in return for mending his father's generator, he was lending us a ten-foot aluminium dinghy and oars. I told him how marvellous it would be to have a boat.

Giving him an old-fashioned look, Lucy asked: 'Titom, have you time to help me get some green coconuts? My husband says he is too busy with engines.' Turning to me, she added: 'You don't mind, do you, darling?'

'No,' I said, and meant it. 'I'm happy enough.'

It was some hours later that she came rushing into the shed where I was working, her words tumbling out even faster than usual.

She knew she'd been away a terribly long time – there were no nice coconuts on the nearby trees – they had tramped right across to the other side of the island, but they were too tall, as I knew, so they had gone to the ones on the southern shore . . .

'Where's Titom?' I cut her short.

'He's had to go,' she said, hastily. 'He's gone.'

'Couldn't face me, eh?'

'What do you mean by that?' she flared.

'Come off it, Lucy,' I said, 'you know perfectly well what I mean. For your own good, though, I'll tell you that in many places in the world today, a black man still derives a certain status from fucking a white woman. It doesn't matter if she's attractive or looks like Whistler's mother, just so long as she's white. So don't get any high faluting ideas that you are some irresistible femme fatale.':

'But nothing happened!' she wailed. 'Nothing happened! You've got to believe me!'

'Why?' I asked, quietly. 'I don't care a monkey's toss what you do. Don't you realise that?' I carried on with inserting reed valves back into the outboard.

'Please listen to me, Gerald,' she insisted. 'And please look at me. I have never, never, been unfaithful to you. Don't turn away! You are going to hear what happened today, whether you like it or not. Titom did try to make love to me. He said that you were too old a bloke for me and that the sight of me gave him a hard-on. I told him that I was married to you, but he insisted that you would never know if we just sank down into the grass and did it. His shorts were bulging but I told him "No – definitely no." He just shrugged the same way he did when you beat him at arm wrestling – and he apologised. That's the truth.'

'Lucy,' I said, 'I can't believe you. I really can't. And it just doesn't matter any more. You've hurt me terribly and all I want to do now is forget it and get over it. So please, just leave me alone.'

Five days later, Titom towed over his father's dinghy. He was quite amiable to me and he ignored Lucy. That same day, Ronald arrived with a tin of flour and a packet of dried yeast.

'A present from the nurses,' he said, and handed Lucy a note from Janine. It said that Lucy's urine test was negative. Janine had also added a message to us both: 'Just hang in there, you two. We're really proud of what you are doing.'

Eight

WITH MY constant ability to share, it was inevitable, I suppose, that the dinghy was instantly 'ours', that Lucy accompanied me whenever I put to sea, and that she should be the one to name the craft 'Isobel'. Also she insisted on doing her share of the rowing.

Just how treacherous were the currents, wind, rocks and reefs of the Torres Strait I discovered when we ventured out about two hundred and fifty metres to a large rock islet standing off the camp's shore. For a week previously, we had confined our rowing to going round Crocodile Bay and along the shore to an even better fishing rock that Ronald had pointed out to us and which proved lucrative for coral trout, rock cod and even barracuda.

Going across to the islet was no trouble at all, and we landed in a small, sandy cove where we beached the boat. Fishing from the islet itself, we caught some excellent snapper (sea perch) and promptly named the islet 'Snapper Rock'.

When we launched the boat to return to Tuin, however, the wind suddenly blew more strongly and the waves became higher. Sitting side by side, each pulling on an oar, we exited from the cove and were instantly swept backwards towards jagged rocks and reefs. Lucy was very calm as we warded them off with our oars, the dinghy spinning and weaving, though I did call her a few different names on two occasions when she pushed the wrong way. Clear of rocks and reefs, we were heading for the Arafura Sea, next stop India! We had, however, swung round into the lee of Snapper Rock and had only the current to combat. But although we pulled with all the force our bodies could exert, we couldn't make headway. It was one of the few occasions when I wished I had someone with me who was as strong as I was.

But before we would reach the open sea, we had to pass through a row of three islands. Our only hope was to guide the dinghy to one of those shores.

Then something occurred which we later came to know very well – the current suddenly changed to run in completely the opposite direction! This takes place every six hours in the Strait, the waters flowing either to the Arafura Sea on the west, or to the Coral Sea on the east.

The dinghy spun round and we were making for the rocky shore of Snapper Rock on the side furthest from Tuin. The aluminium bottom scraped a few times over submerged rocks before the water was shallow enough for us to jump out and haul the boat in.

'We'll just have to wait until the tide drops and we can tow the damn thing back,' I said. Naturally, Sod's Law decreed that on that particular day, the tide didn't drop to its normal level, another peculiarity of the Strait's waters.

We were both completely naked and had to hide when a dinghy went speeding by from Badu. Also, with the wind and spray from the rocks – and the sun behind a rainless cloud – we soon became extremely cold. I thought, then, how silly it was that our relationship was such that we couldn't cuddle for warmth. 'We'd better do exercises,' I said, sharply, 'or we'll lose our body heat.' The temperature had probably dropped to below eighty degrees Fahrenheit. In the shelter of a large rock, we jumped up and down in rhythm and did physical jerks.

Another half-hour or so and I could see that the tide was not getting any lower. 'You sit in the boat and row,' I told her, 'and I'll tow you.'

'You can't,' she countered. 'It's too deep and with the waves you'll never be able to see any sharks.'

'We can't stay here and freeze,' I said. 'Besides, I'm starving.'

Off we set, the water rising up my body until, at its deepest, the waves were hitting my chest. I heard a splash.

Once more, Lucy was by my side, pulling on the rope with me. 'As I told you,' she said, 'on Tuin, I am your woman and will always stand by you.'

'I don't think you have jumped out of the boat because of that,' I said, evenly, 'it's simply that you can't bear to be under an obligation to anyone – especially to me.'

Always after that we took clothes, which were also very necessary to put over the aluminium seats to stop the hot metal from burning our bottoms! Lucy was a good sailor and never felt seasick no matter how long we rode at anchor, bobbing up and down as we fished.

It was essential that we also took with us a bailing tin – not for that specific purpose, though, but to pee in! It was much more convenient and safe than trying to do it over the side.

Once as she sat facing me, holding the tin beneath her open thighs, her stream thundering into it, she said: 'Dah-ling, you *must* write about this in your book. No one ever describes that sort of thing. In stories about lifeboats, and men, women and children surviving for days and weeks on rafts, no one ever has a shit or a piss. It's ridiculous.'

Holding myself because I was bursting and waiting for the tin, I said: 'I know how I'll describe you – "She pissed as long and as loud as Flanders Mare!" ' She guffawed at that.

Because of the deceptiveness of the wind and the deceitfulness of the current, we were twice more stranded on Snapper Rock. The first time we were towed back by some mullet fishermen; the second time by the white headmaster from Badu who had chanced on that particular day to bring two curious Australian school inspectors to see us, after they'd heard on the radio about our survival project.

We went with them in the headmaster's twenty-footer to show them a good fishing spot for large blues that Ronald had taken us to in his dinghy, but we hadn't been able to reach again using only oars.

Lucy never stopped talking. I was used to it, of course, but there was one point where even I was dumbfounded.

Lucy was flitting from subject to topic to subject. As though it were part of the sentence she had interrupted, she said: 'Oh, look, I've caught a wobbegong!' The sleek, beautiful, leopard-skin shark, some four or five feet in length, swam lazily on the surface. 'Of course, my husband and I almost died when the water became salinated and we didn't notice it,' she continued, apparently oblivious to the shark she was pulling closer to the boat. The three men sat stock still in amazement. 'If it hadn't been for Ronald Lui, of Badu, we would have . . .' The shark suddenly spat her hook clean out of its mouth. 'Oh, dear, I've lost it – we would have died. But Tuin is such a beautiful, friendly island – we didn't really want the shark, did we, darling? – and we never feel afraid – look! I've still got my bait! – we have lizards as camp pets. We thought about eating them once . . .'

One of the school inspectors sidled closer to me. 'How on earth do you stand it?' he whispered sympathetically.

'I don't know,' I said. 'Actually, I think she's a wonderful person – and I'm very, very proud of her.'

'Yes,' he mused, 'I suppose she is quite a remarkable woman.' With which I could only agree.

Two days later, after hearing about our being stranded on Snapper Rock, Ronald dumped a rusty Johnson six h.p. outboard on the beach. 'If you can make it go,' he said, 'use it on your boat. And here's half a tank of petrol as well.'

For three days I worked on that engine. Lucy helped by cutting gaskets out of cardboard and looking to see if sparking occurred across the two plugs' gaps when I pulled the starter cord.

On the fourth morning, after running the engine at anchor, I announced, proudly: 'OK, Lucy. Ready for the pleasure trip!' Halfway across Crocodile Bay, the petrol hose connection, which was badly worn and leaking, came apart. We rowed back. 'What a lovely pleasure trip, darling!' But she said it with humour, not sarcasm.

Always with other people's outboards that I later repaired, I would invite Lucy on the 'pleasure trip' for testing. Several times we had to row back because of a minor fault, and it became a standing joke between us that whenever I announced a pleasure trip, she would pretend to groan and say: 'Not *another* exciting pleasure trip, darling?' But she would be in the boat before me, sitting all expectant, prim, nude and radiant.

She bestowed the name 'Jemima' on the small engine. After Ronald brought us another petrol hose, Jemima never once let us down, even when facing fierce currents and waves that drenched us. We went all round Tuin and picnicked on the nearest uninhabited islands. The more we were tossed high by the waves the more she loved it, and constantly she had to use the bailing tin for its correct function.

Although the maximum permitted horsepower to propel a ten-foot dinghy is only ten, I had no alternative but to put much larger engines on the back of Isobel for testing. One of them was a forty h.p. Johnson! I barely opened the throttle and we almost left the water. I'm sure at one stage we touched fifty knots!

'Gosh! That was almost orgasmic, darling,' Lucy gasped, as I cruised the dinghy back and through the only channel between the rocks to the camp beach.

Ronald, who now kept a jar of sugar in the shed for his tea during his frequent and lengthy visits, asked me one morning if I would like a shotgun. Naturally I said yes. 'It's a bit dirty,' he said. 'Don't worry about that,' I replied, 'I can clean it.' Then he told me just how dirty it was. For many months it had leant against the wall in a small room in his house beneath where a python slept. Pythons were not uncommon in houses on Badu because they ate the rats! 'The snake has been pissing and shitting down the barrel all the time,' he said.

When he brought the gun across a few days later, dirty was decidedly an understatement. A single-barrelled twelve-bore, it was covered with thick rust and the barrel was, indeed, almost full of dried python shit. The hammer and trigger were immovable and the gun wouldn't break at the breech. 'Here are three cartridges,' said Ronald, 'just in case you do get it to fire.'

I began working on it straight away with scraper, petrol and oil to get the wooden stock off so that I could break the gun and push a stick down the barrel.

Invariably, whenever Ronald wanted to know something, he would introduce his question by the statement: 'Now, you are an educated man,' then go on with: 'Can you tell me . . .?'

As I sat, working on the gun, Ronald went through his little preamble, continuing with: 'What you think of *puri-puri*?'

'*Puri-puri*?' I asked. 'What's that? A soup or fruit or something?'

'It's about these men who have strange powers,' he said. 'Like they do in Africa.'

'You mean witchdoctors?'

'Yes, that's it – we call them *puri-puri* men. What do you think about them?'

Taken completely off my guard, I could only reply: 'Well, not a lot at the moment. I don't think I would ever want, actually, to challenge one, even if I didn't strictly believe in what he could do. Something I've always steered very clear of. Perhaps my will-power wouldn't be strong enough. Why?'

Ronald had nodded his head in agreement as I had spoken. 'That's what I think,' he said. 'I don't want to believe in it and I don't want to have anything to do with it. I tell everyone on Badu to leave it alone.'

'You mean it still exists there?'

Once more he nodded. 'It did until a few months ago,' he said, 'and now no one really knows what to believe. Let me tell you about it.'

For several years a man on Badu had claimed he had the powers of *puri-puri*. Several unexplained deaths had occurred on the island which could be linked to or were coincidental with arguments or rows with the *puri-puri* man. (I will call him 'P' because, so far as I know, he is still alive.) He never told an intended victim directly

232

that he or she would die. It was always by inference or innuendo. Sure enough, though, that person died shortly afterwards.

It was the last death, only a few months before our arrival, that caused a public outcry, for it involved a twelve-year-old boy who had often given cheek to 'P'. One morning after the boy had cheeked him, 'P', in front of witnesses, had said: 'You will never insult me any more.'

An hour later, the boy was swimming close to the hull of an anchored cargo boat when the captain ordered reverse thrust to take the bow off the sand. The boy was sucked through the large propeller and chopped to pieces. Although the white Australian police were called in, nothing could be proved against 'P' that would stand up in court.

'So what happened?' I asked.

'He was banished from Badu and not allowed to live on any of the islands,' said Ronald.

Lucy was quite shocked by the story, especially as Ronald told it with a more gory and graphic description of the boy's death. She did, however, add a touch of humour to another story of *puri-puri* that Ronald told. The delicious wongai plums were coming into season, and their ripening process could be speeded up by picking them and placing them in the hot sun. If eaten before completely ripe they produced severe wind and diarrhoea.

Ronald's father had told him when he was a little boy of another man on Badu who professed to have the powers of *puri-puri* and was often seen sloping off to a secret place in the bush. One day, a few of the Islanders decided to follow him and, to their astonishment found him standing on his head, completely naked, with his legs in the air and flames shooting out of his backside.

Remembering a particularly nasty eruption she had recently experienced, Lucy said, impulsively: 'He must have been eating wongai!' I roared with laughter and Ronald chuckled heartily as well. Then he became serious again. 'There are now many people on Badu who do not believe in *puri-puri* but I have never heard any of them say, openly, that it is bunkum,' said Ronald. 'But, of course, there are many, especially the older ones, who say that *puri-puri* is very real and very powerful.'

'What about you, Ronald?' I wanted to know. 'What do you say?'

He regarded me for a moment with his blood-red eyes. 'I think, probably, that like a lot of people, I am too afraid to say anything.'

It was my turn to nod. 'I think that's the real answer to it,' I said, 'fear. Especially fear of the unknown. That, quite likely, is what gives a witchdoctor his power over others. I'll tell you one thing, Ronald,' I added, chuckling, 'I would hate to be called upon to mend a *puri-puri* man's engine and find I couldn't get it to go!'

'You'd be dead within a day, that's for sure,' said Ronald, solemnly.

We talked about *puri-puri* for quite a while longer; and though I put forward a lot of opinions and spoke of induced illness, nothing by way of a solution or comfort could be imparted to Ronald. The superstition ingrained in him as a child and young man was too strong for modern-day reasoning.

It was two days before the shotgun was ready for testing. I tied it to a tree, the muzzle pointing at one of the many giant anthills that could be seen all over Tuin. Some of them were above head height. Tying a long line to the trigger, I played it back to the creek, where Lucy and I lay flat on our stomachs. I pulled the line – and the gun, hammer cocked, fell out of the tree!

Securing it to the tree in the same position, I had another go. The blast took the top off the anthill and the barrel didn't burst.

'Let's go hunting!' I said. I knocked a Torres Strait pigeon out of a tree with the first shot, and sent a fast-running scrubfowl arse over tit with the second.

'My hunter! My hero! My husband!' gushed Lucy, like a heroine in *Ripping Yarns*. I thought she was surely taking the piss, but she wasn't. To her, what I had done was miraculous!

It became clear to me, then, that either she hadn't believed me when I told her of my experience with guns or she had been too uninterested to listen. It must have been the latter, considering an incident that had occurred on TI. I was invited to take part in the local clay pigeon club shoot. Lucy decided not to accompany me.

I was feeling rather wretched about our relationship and handled the shotgun like an idiot, bringing a few scornful chuckles. When the final shoot-off came, however, I told myself: 'Come on, Kingsland, get a grip of your knickers.' As I picked up the loaned twelve-bore, I felt the calmness descend that is so very necessary when using a gun or rifle. Ten times, I called 'Pull!' and ten times the clay shattered.

Naturally, like a small boy wanting to boast to his mother about his accomplishment, I couldn't wait to tell Lucy. 'Yes, very good, darling,' she said, and that was all.

Many convolutions of thought went through my mind as we carried the dead birds back to camp. Perhaps she had always regarded me with the same suspicion as I, as a proud wearer of the Red Beret, had looked upon down and outs in London pubs who bummed drinks off me, declaring that they had dropped at Arnhem. After all, in her mind, perhaps I was a down and out. I had nothing to prove that what I had done and been was in any way true. Not being a materialistic person, I had never bothered to file or keep any of the many magazines and newspapers in which my name had appeared.

Also, of course, with my liberal use of Anglo-Saxon words and rough and ready attitude, it was extremely difficult for her to accept that I had once been the focus of attention of a board of directors, spoken to Prince Philip and even offered Racquel Welch a job after having a few drinks with her in the Playboy Club!

The pigeon was tender enough, but although Lucy boiled and boiled the scrubfowl some of the meat was impossible to chew. But we sucked the last remnant of deliciousness out of every piece.

Lucy suddenly clapped her hand to her mouth, exclaiming: 'Oh, my God, Gerald! I've swallowed a shotgun pellet! Is it dangerous?'

'Not to you,' I said, keeping a straight face. 'Only to me if your arse is pointing in my direction when you fart tomorrow morning!' It took her a full half-minute before she began laughing, her whole body shaking with mirth.

I decided, though, never to fire the gun again and declined Ronald's offer of more cartridges. The barrel pittings, inside and out, were far too deep for safety.

'I think you should have a rifle here,' said Ronald, and he brought me his .22 and a box of cartridges. Somehow, I didn't think it right to shoot anything else on Tuin after I had shot one more scrubfowl. The creatures were all part of the island, as we had now become. We had plenty of food coming in and never, since I left Korea, have I shot anything for pleasure. To her joy, I taught Lucy to shoot, and she became a first class shot, putting close groupings in empty petrol cans.

Just as one never appreciates the day-by-day growth of a child, kitten, puppy or plant, so I never really noticed the gradual development of Lucy's voluptuous hour-glass figure. Not until Janine and Bunty paid us a surprise visit.

They had jointly purchased a new fourteen-and-a-half-foot aluminium dinghy and thirty h.p. outboard. Down wind of us and with an Islander guide, they rounded the north-western point and caught Lucy and me nude, up to our thighs in water as I was trying to spear a shovel-nosed ray.

'My God!' exclaimed Bunty, in a very loud voice, as I sank my loins beneath the surface. 'Lucy's got boobs!'

It was then I looked at and appreciated the metamorphosis from skinny, wrinkled flesh to luscious, fleshy curves. The nurses, who didn't come ashore as they were eager to go cruising, were astonished and pleased with our appearances.

'I simply can't believe the difference,' said Janine. 'You,

Gerald, well, you're not the same person. You really have got a fine, muscular body for a man your age.'

With their dinghy speeding away, I raised myself out of the water and walked over to Lucy. 'You know we are very close now,' I said. 'We never row. In fact, we are like a loving husband and wife in every respect but one. Did you know that you really are gorgeous? Stunning! I am not saying now, right at this minute, though I would love to, but it's going to be only natural that I'll want to make love to you.'

'Why don't we try cuddling first or even touching?' she said.

But, with the long deprivation, I needed the sanctimonious severity of a saint not to feel aroused on touching that sensuous, swelling-curved piece of pulchritude with its beautiful golden tan that extended even to under her full breasts and high up between her soft thighs. As soon as she felt my arousal she withdrew from me.

Over the next two weeks, we found ourselves touching each other more and more. A good excuse for this was her suggestion that I taught her the waltz, when I'd discovered in conversation that she had never learned ballroom dancing.

'Is it always necessary to dance as close as this?' she would ask.

'Oh, yes,' I would reply as, to the probable amazement of Tuin residents, we would glide over the sand to the only music we had – my call of one, two, three! Always the impatient frustration within me was unbearable, and several times I angrily accused her of being a 'sadistic, cruel bitch'.

Suddenly, Lucy became broody and sulky, not at all like her. 'What is the matter?' I asked, thinking I had been too harsh in my constant sexual approach.

'I am not *doing* anything,' she complained.

'I know,' I said. 'I wish you would – especially in the tent at night!'

She gave me a withering look. 'Don't you see? It's *you* who provides all the food now with your work on engines. I'm left out of it.'

'Don't be silly,' I said. 'You even go fishing on your own, and you do all the cooking. Never do I have to ask for a cup of tea or a meal. It's always ready for me.'

'You don't understand,' she said.

'I'm not that thick,' I replied. 'I think I do. Look, it's well into September and the feature we promised the *Sunday Telegraph* is a month overdue. Why don't you write it? I'm far too busy with engines to even think about it.'

'You're the writer,' she said. 'You do it.' Later, she asked: 'Do you really think I could write it?'

'Positive,' I said. 'Your diary is beautifully written. I only wish I could write like you. You have a fantastic way with words. Just set it down the way you write to your mother and father. Be natural. Incorporate excerpts from your diary. It's simple. Just don't try to be clever.'

Her first effort was, as I expected, much too forced. I gave her constructive criticism and scribbled a few paragraphs as guides. The second attempt, after I'd suggested she altered one or two things and pointed out a few spelling mistakes, was very good. I lent her my typewriter for neat presentation and she spent an hour a day, typing with two fingers.

It was then, without giving me any reason, she returned one afternoon from the other side of the island and said, simply: 'I have made an important decision that you are going to like very much.'

That night, she tried to give herself to me but she was far too tense. 'It's all right, darling,' I told her, after she had apologised, 'we have all the time in the world.'

The following night saw the beginning of months and months of paradise for me. The next day, Lucy finished the *Telegraph* article with: 'We have sun, sea, sand and sex – we are very happy!'

Once again, Lucy decided to give a name to something – my now over-worked appendage. She called it the 'Member for Tuin West'! Often she would lead him down the beach and we would frolic in the sea, washing each other and caressing without actually making love.

Once we did attempt to, but the sight of a shark's fin quickly put paid to that!

Humour, jokes and quips flowed between us all the time. She was very proud of her new-found breasts. A favourite act was to cross her eyes, point to one breast and say, in an idiot's voice: 'I've got two of these. One! Two!' Sometimes, she would approach me, coquettishly, as I was bending over an engine, and say, also in the voice of an idiot: 'Do you live on Tuin? I do. Would you like to make love to me?'

'Piss off!' I'd say, laughing. 'Can't you see I'm busy.' But always, the Member for Tuin West insisted on retaining his seat and pleasing his constituency. Knowing him very well, Lucy would quickly spread my large green towel on the sand and we would sink down upon it.

'It's no wonder that the cheeks of my arse are getting extraordinarily tanned!' I commented on one occasion.

'Oh, you poor old darling,' she soothed. 'Are they getting burnt? Let me cover them with my legs . . .'

Nine

CONTRARY TO what Lucy persistently maintained, I was firmly convinced that I was no longer being Robinson Crusoe. She argued that the survival project was still very much alive, for we were surviving in an alien place, and it didn't matter that our surviving was due to my skill with engines.

'What difference is there,' she pointed out, 'between using a spanner and using a garden spade and fishing line? And it certainly isn't our fault that we were put on an island that cannot sustain life all year round.'

Both the creek and the well – even though I had dug down another couple of feet in the latter – had gone completely dry. All the vegetable plants had shrivelled and died, even though there were, now, fairly heavy dews in the mornings. The sun was directly overhead again, swinging its way to the Capricorn line for Australia's summer. I put in a few more seeds in the place where the earthworms were – though they had gone to a lower level – but not one shoot dared to show itself to that sun. It was as though the ground were completely dead. Two weeks later I uncovered the seeds. They were as dry as tinder and intact.

What Lucy said made sense, but nothing could change the fact that, in my mind, I knew I wasn't living the character role in a book I had set out to write. I was, however, formulating another idea. '*You* are the real story,' I told Lucy. 'No one is interested in me, but they would be in hearing about a delicately wristed former Inland Revenue clerk surviving in these conditions. It's a natural best-seller: the first Mrs Robinson Crusoe!' She seemed quite pleased with that.

'The number of visitors that we are getting,' I went on, 'is also making my project a bit of a farce.' Ronald was bringing his children over for day-long picnics or to take

240

us trolling for the giant and mouth-watering queen fish; Islanders were bringing their engines; and another regular visitor was Ronald's giant of a nephew, Ronnie Nomoa, who was usually accompanied by his equally large wife, Enid, and their two-year-old son Charles Torres.

During the last days of September, we had another visitor – and quite a surprise it was, too. Lucy and I were out fishing in Isobel when we saw the large dinghy heading for our camp beach. There were white men in it and, as we drew closer, we recognised one of the two yachtsmen, Peter! It was with very mixed feelings that I shook hands with him. Lucy, too, was at first reserved. But over a cup of tea, his and Derek's previous visit was forgotten except that Peter repeated over and over again: 'I can't believe how well you two look. And your happiness shows. A lot must have happened since we were here.'

'It certainly has,' said Lucy, smiling and moving closer to me.

'What happened to you, then, Peter?' I asked. 'And why are you here?' I knew why, for we had received a letter from him and Derek a month previously, saying that they had been forced to cancel their voyage when their craft had been impounded at Irian Jaya, a short way along the New Guinea coast.

'Didn't you get our letter saying we thought it better that you didn't come back to Tuin?' I further asked.

'Yes,' he said, 'but as I wrote, we are here to make a publicity film, re-enacting our trip and we would like to include you two in it.'

After a brief word with Lucy, I said: 'If you want to take some film of us, you can. Any publicity can only be good for us.'

They returned next morning – they were on a boat anchored off Badu – and filmed Lucy and me for about an hour. Ronald had long been asking me to go to Badu and wire his house for a new generator. I decided to give him a surprise, and went with Peter and the camera crew to Badu, knowing that Ronald would bring me back that evening.

Lucy refused to accompany me and said that my going

241

was tantamount to throwing in the towel where the survival project was concerned. I argued with her that from my intended book's point of view, Tuin had become another failure when all the people started to arrive.

'It isn't a failure to me,' she said. 'I don't want to set foot on an inhabited island until at least Christmas. I'm sorry, but that's the way I feel and that's what I want to do, if only to prove something to myself.'

By my going to Badu, yet another chapter in the Torres Strait episode was added – for, after Ronald had shown me the way through the reefs and around three very large and jagged mounds of submerged rock, I frequently made the journey on my own in *Isobel*.

Whenever I asked Lucy to go with me, she would reply: 'Can't you see – here, on Tuin, I am *me* – being what I decided to be and do for one year.'

'I think you are just playing a stubborn, make-believe game of trying to be a castaway,' I told her.

One of the major tasks I had to do on Badu was to free the clutch on Crossfield's beloved tractor, 'Red Rose'. He had never been able to use it since he had bought it a few months previously from the company which had built the airstrip a few miles from the village. He had made the purchase after the tractor had been loaded on to the company's ship and he had requested it be 'dropped off' at the village.

The crew had done just that – into about four feet of water! And they left it there! It was too dark to do anything about it, and it wasn't until the following morning that Red Rose was towed out, her clutch seized solid, for the tide had been out, in and out again.

With an audience of about twenty-five wide-eyed Islanders, I split the tractor in half with a jack and block and tackle and removed the clutch from the flywheel. Everybody wanted to help to clean it! The clutch was like a new pin when I re-assembled it. Then I rejoined the two halves of the tractor.

Crossfield was overjoyed at seeing his proud possession working at long last. 'You bloody good mechanic,' he said.

'You look after all my engines. What you want from store? Come, I buy for you, Mister *Governor* of Tuin.' And he laughed loudly.

Always, the Islanders were generous with their gifts of food – for I would never take money – and I would return with Isobel laden, my heart full of love and impatient excitement within me to see Lucy again after being away for only a few hours.

One particular evening, when I had a special surprise for her – apples! – I was later than usual in getting away. In Isobel, the journey took between forty minutes and an hour, depending on which way the current was running. The vicious cross-currents around Wia Island were worse than I had encountered before, and the waves were alarmingly high. Jemima never so much as coughed with the extra exertion, though Isobel was stopped dead in her tracks several times and I was completely drenched. With the hot sun drying me, my body was white with salt by the time I got through.

Lucy came down the beach the minute she heard the engine, and waded out until she was knee-deep in the sea. The low sun flamed everything and she looked like a bronze-gold Helen of Troy, her breasts full and pointed, her narrow waist flowing out to rounded hips and long thighs, her upside-down triangle of pubic hair curling silvery brown and soft. I saw, too, that she was wearing her sun-drenched hair in one long tress over the front of her right shoulder as she had on our wedding day.

'I was getting worried about you,' she said, simply. As I gave her the apples and heard her shriek of delight, I considered myself to be one of the luckiest, happiest men alive. . .

That night she asked me: 'When are you going to build me a house like the one you and the boys built on Cocos?'

Ronald brought me hammer, saw, nails and spirit level, and told me I could have the table in the shed and borrow a giant sheet of yellow canvas for one side. Crossfield gave me boards, cut pieces of timber and a large roll of plastic sheeting. I designed the house long and narrow, like a ship, making the width the nine-foot length of the corrugated iron sheets for economy and facility.

I put love as well as sweat into that house, making a small breakfast table and working surfaces in the kitchen section; I made another small table for the 'living room', and a window of plastic sheeting stretched over a wooden frame that could be opened and closed. At the far end was the bedroom where I made a platform of smooth boarding at knee-height for the inner section of the tent to stand on as a ready-made mosquito net.

Ronnie gave us an old double quilt that could be put under the tent to make a softer bed, and he helped me to put on the roof. We spread the roll of plastic over the entire length, for many of the iron sheets were holed, then placed layers of palm leaves on top to stop the heat coming through. He made a special trip in his dinghy to Moa Island to cut extremely long bamboos, and these we placed on the palms and secured with wire. No water or heat ever came through that roof.

Lucy's kitchen, just inside the entrance, consisted of two converted petrol drums raised so she didn't have to stoop overmuch, a wood store and a Dutch oven given to us by Ronald. As a finishing touch, Lucy and I went in Isobel to a section of Palm Beach, returning with a load of minute white sea shells which we spread over the sandy floor to make a carpet. An extra were lengths of rush matting which were included in one payment for engine repair.

Surveying the finished work, Lucy kissed me and said: 'No one has ever built me a home before. I feel like a proper wife, now.'

October brought the mango fruit and the *naigai* season when the South-East Trades stopped blowing for a month or more as expected. Our navels filled and overflowed with sweat as we lay in the tent at night, convinced there was no oxygen left in the air. Mosquitoes began biting in the daytime, too. There was no respite from the terrible heat – even the sea was hot. Love-making was almost unbearable, with trapped pools of sweat between our slippery bodies.

When I dug up the potato-like cassava roots, which Ronald had pointed out to me, the soil was hot even at a depth of nine inches. The sea looked as if it had died, for it was as smooth as a mirror without so much as a ripple. Going into the deep channels off Wia Island we could look over the side of Isobel and see clear to the bottom a hundred to a hundred and fifty feet below! It was like looking into a new, vast world with shoals of different fish at varying levels and in their several places.

A terrible stench added to our general discomfort one day with the invasion of our waters by millions of small, brown jellyfish. Washed ashore, they roasted and died in their thousands, and their bodies quickly putrefied. Although they didn't sting in the water, they did so after death.

Then, as though the ground and the air were not hot enough, Ronald and Lucy set fire to the entire island! We had gone with Ronald into goanna country behind Coconut Alley to survey the bush for melon and pumpkin gardens in readiness for the pending rainy season. The Islander way of preparing ground is to fire it. Small, hot breezes were constantly springing up to fan but not cool us and, like the currents, changed direction suddenly.

Lucy and Ronald stayed behind to burn the places we had selected while I returned to camp to finish an engine that was required urgently.

On their return, Ronald said: 'It's burning nicely. The breeze is right and the fire will be burned out before dark.' Then, after a cup of tea, he left.

Early next morning, I went outside for my customary pee – and saw a wall of flame creeping towards me! 'Lucy!' I yelled. 'Get out of bed, quickly! The fucking island's on fire!'

'What?' she screeched, and came running out.

Crackles and more flames were spreading on our left and right, and the creeping central wall was the entire length of the island's waist. 'We've got to move fast,' I said. 'Lay every spare sheet of iron in a long row on the ground along this side of the hut and I'll cut down those saplings and brush as a fire break.' When she had finished, she joined me to help drag the cut branches away from the hut. 'How much fresh water have we got?' I asked. 'Three,' she said. 'Ronald brought two more yesterday.'

'Right, I'll use one of them to sprinkle the roof. You fetch sea water in the bucket. Any flame that creeps under the corrugated iron, kill it.'

For an hour we worked and kept watch as the persistent flames stretched themselves under the iron sheets, licking out on the other side. When the heart of the fire was at its closest, we thought our naked bodies must surely blister. At last, we pronounced the fire round the house dead, though it was still spreading over the high ground to the north and south.

'I think that calls for a celebration,' I said.

'What's it to be then?' she asked. 'Tea or the green towel?' And she fluttered her eyelashes.

'Oh, both, I think, don't you?'

'Indubitably, Carruthers,' she said. 'Indubitably.'

After a late breakfast, we contemplated the black carpet that was spread over our beautiful little island. In pseudo-County tones I remarked: 'I say, dah-ling, do you think it might be a trifle hotter today?'

Except for their blackened trunks the gums and pandanus, together with the larger trees, appeared to be unharmed. Even the hardwood bushes had weathered the fire, but we could now see light coming through from the other side at ground level.

'I wonder how many roasted snakes and lizards there are?' she mused. 'And what about the goannas?'

'I shouldn't think many got caught,' I said. 'They can move pretty fast and, probably, most of them tucked themselves away in holes and under rocks. After all, the bloody ground's almost as hot as fire every day!'

To get away from Tuin's intense mid-day heat, Lucy would make up a picnic basket and we would put to sea in Isobel in search of a cool breeze as well as to fish. Ronald had given us a trolling line. When the current was running high and the birds were working, we set off for the queen fish spot between two uninhabited islands about a mile off our western shore. When we had passed over the entrance reef, Lucy sat on the central seat, facing me to play out the line as I kept Jemima at quarter throttle. When the line was fully out, I tied the end to the stern and she held it looped round her open hand. I opened the throttle and steered towards the known spot. We never reached it. The line was suddenly torn from Lucy's hand. Isobel not only stopped dead with the tremendous pull but went back a metre or two before the line snapped. Lucy's legs shot up over her shoulders as she went backwards into the bow, bruising her back, arms and thighs.

'Are you all right?' I asked, as she re-seated herself.

'I think so,' she said, rubbing several places. 'I'll have some bruises tomorrow, though. What the devil was it?'

'I don't know,' I said. 'But whatever it was, I'm bloody glad we didn't catch it!'

'Beautiful Jemima,' praised Lucy, 'she never even cut out. She's a very good girl.'

The rains started that year far earlier than the Islanders could remember. Showers came first. Lucy and I danced in the first one, and it was luxurious to wash our hair in fresh water and not feel salt on our skins. Before, Lucy always washed her hair in the sea, only rinsing with our drinking water.

Thick fog and mists came next, bringing stifling heat with the humidity. Drizzle would last for two or three days, often becoming so fine that it was like vapour and all around was the grey light that is so very peculiar to Venice.

I asked Ronald – who was not at all perturbed about the fire – when could we expect the heavy rains. 'Any time, now,' he said, 'though December is the usual month. You'll see the black clouds gathering in the west. Then the wind will start blowing from the north-west. You won't be able to land on this side of the island. You'll have to keep your boat on the other side.'

One morning after I had serviced Ronnie's thirty-five h.p. Tohatsu outboard, he said it was time we tasted the pure water of the spring at Yaza. This was on the other side of Badu, some eight miles from the village, and it had been discovered many years ago by a warrior when he had thrown his spear. 'The story is that when he pulled his spear from the ground,' said Ronnie, 'the water shot high into the air.'

Taking a couple of Ronald's water canisters, Lucy and I went with him in his dinghy with Enid and Charles Torres. The Tohatsu made a noise like an aeroplane engine, and sent the dinghy planing across the water at what seemed like thirty knots.

'My engine no go this fast for a long time,' he said, with a happy smile that showed his beautiful set of strong, white teeth.

We passed close to Death Adder Island and plunged through the turbulent waters off Badu's 'Soldier Point' with hardly any slackening of speed, the dinghy virtually skipping across the tops of the waves and the spray flying high on either side. I could see that Lucy was thrilled to bits.

Soldier Point is so named because, right on the tip of the rock point, stands a small pinnacle which does look like a soldier standing to attention. Much to the delight of Ronnie and Enid, Lucy and I saluted as we sped past.

At the far end of a long, sweeping bay of golden sand, Ronnie timed the closing of the throttle so that the dinghy's bow settled nicely on the beach. We followed him along a short, overgrown path through dense dark-green bushes to the spring, which revealed itself as a crystal-clear pool and the home of a giant black freshwater eel. Unmindful of the pool's denizen, Ronnie stepped into the calf-deep water and filled the canisters.

'Taste,' he said. I did so. 'If this water could be bottled and sold in England,' I said, 'we would make a fortune!' He accepted the compliment with a contented smile.

Instead of heading the dinghy straight back, Ronnie cruised it around a wide group of black rocks that stood above the water. His eyes were constantly searching. Then he saw what he was looking for – a huge brown sting-ray. It streaked away and Ronnie roared after it. In and out of the rocks, narrowly missing them at times, he raced the dinghy after the flitting, sometimes shadowy shape in the shallows until he had puffed it to almost a standstill. Knocking the engine out of gear, Ronnie drove his spear down hard and leapt from the dinghy.

Little Charles Torres clapped his hands with delight as he watched his father struggling in the broiling sandy water to keep the ray pinned down and avoid the threshing, poisonous tail. Gradually the water stopped churning, and Ronnie spoke sharply in his language to Enid. She quickly handed him a long butcher's hook with a wooden handle. Ronnie drove the point between the ray's eyes and withdrew his spear. Then, as though the creature weighed only a few pounds instead of a good half-hundredweight, he lifted it out of the water and into the dinghy. It thudded on the bottom, flapped vigorously for a few moments and laid still.

'King tucker,' said an enormously pleased Ronnie. 'I show you how to cook it when we get to Tuin.'

At a particular time of the year, the liver of the brown sting-ray turns to what the Islanders call 'fat'. This is cut out, placed into a large, dry saucepan and melted down like lard. All the white flesh is taken away from the skin and painstakingly shredded with one's fingers. Then it is washed in the sea and wrung out hard between the hands until it is soft and dry. This is then dropped into the boiling liver fat. My description of the taste? Similar to plaice with a hint of cod liver oil. Both Lucy and I found it just a trifle sickly, though we ate all that was put before us. There was absolutely loads left over, and Ronald took it back to Badu to share amongst family and friends, as is the Islander custom.

During my working visits to Badu, I had partaken of lunches, *en famille*, of green turtle and dugong. The latter, also known as the sea-cow, is a mammal with a face like a hippopotamus and with a long fish's tail and no hind legs. I was told that the story of the mermaid originated with the dugong, for the sea-cow cradles her baby in her short, flipper-like arms. They are a protected species which only the Islanders are allowed to hunt and eat, and this they do with long, beautifully carved, hardwood poles, with a detachable spray of barbs inserted in one end and tied to a long, thin rope. The dugong can measure up to twenty feet in length and almost every piece of it is eaten. Grass on the sea bed is its only food, and there is hardly any difference in looks and taste between a beef steak and a dugong steak. Unfortunately, anyone eating it, especially the fat, for the first time gets the runs! I was barely halfway through my first steak at Ronald's house when I had to leave the table hurriedly.

The dugong can only swim at full speed for some hundred and fifty metres, then it must come up for air. Reefs in the Torres Strait are miles in length and width and when a herd is chased in the shallow waters above a reef, the animals have no deeps in which to dive for safety. One is selected and chased with outboard at full throttle and is soon an exhausted and easy prey. The barbs do not kill the dugong, however, for they penetrate only the tough, leathery skin and never go into the quick covered by a layer of fat several inches deep. The creature's head is held under water and it quickly drowns. The body soon swells with natural gas in the same way as a land animal blows up after death. The inflated body makes the skin very taut and easy to cut.

One night after Lucy and I had been in bed for an hour or so, we heard engines and much shouting from Tuin's northern waters. This went on for some time, then there was silence except for the usual blood-curdling screams of the 'Devil Birds', as Ronald had told us they were called.

'What was that all about?' Lucy and I asked each other.

Next day, Ronald came over and told us. It was not only a tragedy but a complete and utter mystery. One young man, who had a wife and small children, had thrust the barbs into a dugong, and the creature had made off – as was the usual case when the barbs struck home – sending the coiled rope streaking over the side.

'No one knows how it happened,' said Ronald, sadly, 'the white police are still here and they can find no explanation. The rope somehow tied itself round the man's neck – not a loop, but an actual knot. Of course, he was pulled overboard by the force of the dugong and he was dead before the other boat could reach him.'

Lucy and I, naturally, questioned him at length, and we talked about how magicians can tie knots in ropes without letting go of the ends. Then another thought struck me. 'It wasn't *puri-puri*?' I said.

'Not exactly,' he said, hesitantly, 'not on Badu – but one or two of the old people have said that it might be bad luck to have *marakai* (white man) living on Tuin.'

Both Lucy and I protested how preposterous that was, but I could sympathise with the old folk's superstition, and did, in fact, feel not a little uneasy that a young man should have died so suddenly and mysteriously in Tuin's waters.

'The white police said they were coming over to see you,' said Ronald, 'but I will tell them of what you heard and perhaps they won't come.' The police didn't call on us, and although the tragedy stayed in our minds, we heard no more about it.

With primary schools on the islands, and teenagers being sent to the mainland for further education, it was only logic to reason that the old ways, crafts and superstitions would eventually disappear. An example of the failing to hand down skills and knowledge from generation to generation was the hawksbill turtle. It has two glands within its body and if they are not cut out correctly they poison the meat, and anyone who eats it dies.

Only a few old people on Badu knew how to cut the hawksbill turtle, and none of the younger people bothered to learn from them. Not even Ronald or Crossfield knew how to take out the deadly glands.

'Why bother?' they said. 'We have plenty green turtle. But we'll get someone to learn how to do it from the old people.'

The turtle is never killed before it is cut – that is Islander custom. Many dishes are made from it and the meat tastes rather like mutton flavoured with fish. After a week, I found it very acceptable. But I never did acquire a taste for turtle eggs. The white never hardens when boiled, and looks like mucus on a plate, and the fishy flavour is marked.

Ten

THERE WERE several compelling reasons why Lucy and I never spent Christmas and New Year on Tuin: the Islanders' urgent calls on me to repair home generators for Christmas lights; to ensure that outboards would get them to TI for shopping sprees; Crossfield's summoning me to decarbonise and tune his Landrover; Ronald's insistence that we spent Christmas at his house; and a very heart-wringing plea from elderly, toothless but extremely handsome Rita Nona to mend her house generator's engine.

Her husband, Philip, an ex-council chief, had lost a leg through stepping on a stonefish, was almost blind with cataracts, was a diabetic and had severe pulmonary trouble. Without electricity to drive his fans, he lay on his bed in extreme discomfort, gasping for breath and with the sweat pouring from him.

I made Rita's plea my priority. The generator's diesel engine was made in India and new parts were not available. I found two old rusty petrol engines behind the power house and built one good one with parts from both. It took me four days of intense work before the grounds and house were ablaze with lights again and Philip had his fans. Rita had quite a modern second house which was empty and she prepared a large double bedroom in it for Lucy and me.

We thought the deep, double mattress would be luxury, but so used had we become to our own hard bed on Tuin that, on the first night, we had trouble getting to sleep. During one mealtime, I discovered that the sweet potato garden on Tuin was Rita's and, rather shamefacedly, I told her that we had eaten them. 'No worry,' she said, with a wave of her hand, 'what you have done is worth many potato gardens.' She insisted that we slept there for the whole week before Christmas while I saw to other people's engines.

On Christmas Eve and the night of Christmas, Ronald gave up his double bed for us as honoured guests. His seventeen-year-old daughter, Robyn, was home from mainland school and she and Lucy became good friends.

Large feastings and drinking parties were held in many homes on the Christmas Day. Of course, Lucy and I had to go the rounds, groping our way through rain-sodden, frog-croaking grass, and finally making our way back to Ronald's to fall instantly asleep on the bed.

The little amount of clothing that Lucy and I had taken to Tuin – we had left our really good clothes in Brisbane and on TI – was rotten with tropical mould and, in Lucy's case, was much too small for her fuller figure. Ronald had once more come to the rescue, bringing to Tuin a dress and knickers for Lucy and a shirt and trousers for me.

Lucy had sent to England for trinkets for the children and a steel lighter for Ronald with his initials engraved on the case. She received another dress and cotton knickers, and I was given a pair of gingham shorts and a beautiful, hand-stitched man's skirt (*lava-lava*), the latter by Harriet Nona, wife of Walter whose speedboat I had repaired.

At the communal functions and gatherings over the festive days, Crossfield always introduced us as the 'Governor of Tuin and his First Lady'. Privately, he told us: 'You proper island people now.'

I sensed he was a trifle jealous over Ronald's claiming me as his discovery and protégé, and this was confirmed when Crossfield commanded that Lucy and I stay at his very large wooden house. We were allocated our own private room and double bed, opening off the large room where all the family slept on the floor or on mattresses after watching video films, mostly Westerns.

Sometimes, night after night, the same video would be played, and there was one particular favourite, *God's Gun*, which was re-played so many times that, I swear, Lucy and I learned the script! Only a thin partition separated it from our bedroom and we often took advantage of the rowdy, gun-blasting battles. There was one scene in which the hero, played by Lee van Cleef, called 'Draw!' The interim silence until, as we knew, the six-shooters would bang seemed to grow longer each night and we found ourselves waiting in suspended animation for the detonations to begin!

From dawn to dusk, I was working in the intense humidity of the rainy season. No sooner had I repaired one engine than I was called upon to repair another. Many times, after a shower and putting on my *lava-lava*, I would start to nod off as Lucy and I sat eating the evening meal with Crossfield, his enormous, jovial wife, Teleai, and their innumerable family.

'Lucy, what you do to him each night?' said Crossfield, bringing laughs and chuckles from everyone.

'Gerald work too hard on engines,' said Teleai.

'That's for sure,' I said, sighing. 'That's for sure.'

Most of Lucy's time in the day was spent helping me with engines, handing me spanners like a theatre nurse assists a surgeon. One morning she went off with the women for a bathe in the nearby swollen stream, leaving me to work on an engine in Crossfield's garage. She returned, looking radiant in a silky, blue-patterned sarong. 'Wait until you see what I've got underneath,' she whispered.

Like a naughty boy and girl we went behind the garage and she slowly opened her sarong. She was wearing an ultra-sexy pair of pink nylon french knickers. 'Wow!' I said. 'Where did you get those?'

'Millie gave them to me,' she replied. 'Aren't they nice? Can't we go somewhere?'

'Yes,' I said, 'I have to look at the water pump engine. Good old Millie. I always thought she was a sexy piece.' Millie was one of the prettiest girls on Badu and Crossfield's private secretary – though, of course, she couldn't type!

Lucy rode with me on Red Rose out into the bush. When we reached the large river near the pump, she asked

254

me to stop the tractor. She dismounted, stood facing me and did a slow strip-tease until she was wearing only her newly-acquired pink knickers. 'It's really nice to feel something silky around me again,' she said, pulling the knickers taut across her tummy. 'Let's make love in the water.'

'And get my arse bitten by a crocodile?' I said, jokingly. 'Besides, look at that torrent!' So we sank down on to the sparsely-grassed sand and took our chances with the death adders . . .

Twice we tried to return to Tuin, but Isobel and Jemima were no match for the high waves and turbulence whipped up by the north-westerly, and we were forced to turn back.

So overjoyed with the two-fold power his Landrover now had, Crossfield gave me a twelve-foot dinghy and an old Mariner twenty h.p. I worked on that engine assiduously. It was our only hope of getting home, for we had no wish to return without our own boat.

Both Crossfield and Ronald came down to the beach to see us off. 'I'll send a dinghy over if I need you,' said Crossfield. 'Remember what I've taught you, my son, let the boat roll sideways to the waves – don't face them.'

Several times, Crossfield had taken Lucy and me out in his fourteen-and-a-half-foot dinghy, either to fish or to show us other parts of Badu, and I had seen how he rolled the dinghy with the waves.

Just as I was about to pull the Mariner's starter cord, he said: 'By the way – something we have been meaning to ask you – where is your other woman?'

'What other woman?' I said.

He held up two fingers. 'You have two womans with you when you first go Tuin.'

Laughing, I explained about Jackie and her being taken off by helicopter shortly after we had landed.

'Good,' said Crossfield. 'We thought maybe she had died and you buried her.'

Lucy and I thought how incredible it was that they had

waited all that time until asking me about something that had obviously puzzled them.

Our dinghy was laden with a twin-jet paraffin cooker, a gallon of paraffin, an empty forty-five gallon drum for catching rainwater, a .22 rifle with telescopic sights for more target practice, lots of food and a twelve-pack carton of beer.

As soon as we were out of sight of the village, Lucy opened two bottles of beer, handed me one and performed a provocative strip-tease with her only two garments. Then she sat facing me with her legs suggestively apart, licking the bottle neck of her beer with her pointed tongue. 'You wanton hussy,' I said, chuckling with immense pleasure and delight.

Before leaving Badu's lee, she put her knickers on again and draped the sarong around her in readiness for the comparatively chilly wind and spray, although the sun shone burning hot and the temperature was above eighty degrees Fahrenheit.

I did as Crossfield had shown me and, although each wave thudded along the side and drenched us, the dinghy rode easily. The violent lurching and juddering was, however, slightly startling at first, causing Lucy to grip the top of the tumblehome with two hands to stop her from being shot overboard. Because of my tacking to run sideways along the waves the journey took longer than it would have done ordinarily.

As we rounded Wia Island and emerged from the main turbulence, I could see the waves pounding on our camp beach, but we had no desire to go to the other side of Tuin and carry everything across. In any case, the surf was chicken-feed to me after my experience on Cocos and presented little trouble.

'Oh, it's lovely to be home, darling,' Lucy said, after we had unloaded and beached the dinghy. She sat down in the water and rose so that her pink knickers were transparent and clung to the roundness and dips of her contours. The second thing we did was to eat a meal. With a mixture of sadness and happiness, I commented: 'This

256

is how it should have been on our first day on Tuin. What a beautiful love story I could have written.'

We had been away for more than three weeks and I am sure that our lizards, albeit a trifle shy, were pleased to see us. So was 'Mr Poo', our pet caucal pheasant, who visited us frequently for scraps. He came strutting through the camp and performed his customary and ludicrous act of standing upside down in a bush in his search for large insects!

This particular kind of pheasant is called the 'Poo Poo Bird' by Islanders and is a rain caller. Many, many times Lucy and I had heard his plaintive calling of 'Poo Poo' for rain. Now, at long last, his prayer was constantly answered.

The rainy season had brought to Lucy and me another bugbear – tinea, or footrot. But it was painless although the raw areas looked revoltingly sore. Every evening we had to wash, dry and powder our feet before getting into the tent.

Our re-found paradise on Tuin was interrupted after only three weeks with the unexpected arrival one morning of Ronald, almost catching us *in flagrante delicto* on the green towel. Philip Nona had died suddenly that morning, and Rita wanted us to pay our respects.

We arrived to a house of mourning and were silently led to a large room where Philip's body, completely wrapped in a white sheet, lay on a trestle table, the family sitting around it on the floor with their backs against the wall. Islanders show respect for living or dead by presence: the giving up of one's time and according it to another. With not a word being spoken, just nods of recognition and acceptance, Lucy and I sat together close to the body and gave Philip our presence for fifteen minutes. Then we rose and silently left the room.

For lesser people who had died, coffins were made on Badu. But Philip had been a highly respected and well-loved chief and an ornate coffin had been ordered from the mainland. For two days until it arrived, the corpse was kept in the community freezer with the crayfish. Lucy and I could do no other but stay on Badu for the funeral.

After a service in the church – founded by what the Islanders called the 'Coming of the Light' (the London Missionary Society) – the coffin, with the village priest Father Blanket and the black-dressed huddle of the family leading the trailing mourners, was borne on the island truck to the cemetery.

Bed-sheet upon bed-sheet was laid in the grave before the coffin was lowered by Philip's sons, their eyes filled with tears. So that no earth should touch the coffin, more bed-sheets were placed on top and around it.

Personal gifts, such as hand-stitched lava-lavas and new shirts in plastic wrappers were next laid in the grave, and overall was spread a quilt.

To a cacophony of wailing by Rita, her daughters and other womenfolk, Philip's sons and their cousins rapidly filled in the grave, the last shovelfuls falling with the darkness. At the feasting and drinking which followed, even Father Blanket was legless!

A feeling of exalted warmth spread through me when Rita approached me and said: 'We will always be grateful to you for making Philip's last Christmas a happier one.'

If funerals were excellent excuses for merry-making, a tombstone ceremony was even better. Lucy and I attended two. The first, before Christmas, was for a deceased member of the Nomoa family. The second, just before we had gone back to Tuin, was for the father of Teleai, and was the more spectacular because he had been the last real chieftain to hold sway over the Australian government, and he had hated all white people.

Several years always elapsed between burial and erection of a tombstone because the latter cost thousands of dollars, and the family had to save very hard. For the tombstone ceremony of Teleai's father, nine pigs were slaughtered with the knife, three giant turtles were hunted and cut up alive and two full-grown dugong were also caught. Large *kapmaurie* fires were always built for feastings – all meat and bread wrapped first in banana leaves then in palm leaves and covered with hot stones.

A tradition of the ceremony is island dancing in separate groups of men and women, all wearing costume and paint and accompanied by the chanting of songs in Badu language, the men brandishing spears and bows and arrows.

Lucy and I sat enraptured and entranced the first time we saw it. I had made the night-time dancing more spectacular by wiring up four spotlights which glistened the painted faces and glinted the weapons, emphasising the paganism of the dance. Rhythm came from the beat of a stick on an empty petrol can, after the tribal drum suddenly broke, and the lilting call of a song-leader.

Folklore forbade any form of activity on the day following a funeral and any boat that put to sea was said to be doomed. So Lucy and I spent yet another night on Badu. Sod's Law once more took a hand, for that night the real violence of the north-westerly hit the Strait, and blowed on and off for several weeks. Only the very big dinghies with powerful engines put to sea, and only when necessary.

That was how it came about that Lucy spent her twenty-sixth birthday on Badu and I spent my fifty-second there. For my birthday, she baked a surprise cake and Crossfield contributed fruit. For some reason, Lucy insisted I told no one about her birthday until it was over.

At the times when the sea was navigable by our small dinghy I always seemed to be in the middle of some engine repair. Also, much to Lucy's discontent, I was quite enjoying life on Badu even though I did want to be alone with her on Tuin.

'I just hate being a mechanic's missus and walking in your shadow,' she complained, frequently. 'You are king here – but what am I?'

'You are Lucy Kingsland,' I told her, 'and everyone thinks very highly of you. The children absolutely adore you. Can't you see that this is just one of life's many eras? Nothing is for ever, and I certainly don't want to be a mechanic for the rest of my life and I don't intend to. My true profession is publishing and journalism. You know that – or you should do, now.'

'Then when are you going to start writing your book?' she retorted.

'I don't know, Lucy. I might be able to get some angle to it all if I knew what you were going to do – stay or leave. For instance, if you stayed with me, I would work here for money, buy a big boat and we could go right across the Pacific visiting all those little-known islands – even mending some engines en route. We'd have a wonderful life, we could both write and the world would be our oyster.'

'I think you should write this book first before thinking of other adventures,' she said dampeningly.

Although Crossfield, Teleai and their closest friends often referred to Lucy and me as the 'Lovebirds', some of the more ignorant and older people, living on the outskirts of the village, seemed to regard me as being something of a *puri-puri* man and some used me to get their children to obey. 'You come here,' a mother would shout to her disobedient child, 'or *marakai* eat you!' That always sent the little sod screaming to her skirts. Even if Lucy tried to pick up or even touch some of the very small toddlers, they would scream blue murder and run like the wind. I only had to look at some of them to get the same response.

The older people's attitude towards me was understandable because some of their engines that I resuscitated were little more than rusty heaps of iron that had lain dead for months. Some, in fact, had been quickly recovered from the local dump when the owners had heard of my skills and cheap labour.

Many times the delighted owner, on hearing of his or her heap of rust running smoothly after a few initial coughs, would accuse me: 'You put *puri-puri* on that engine!' And nothing I said would convince them otherwise.

Some of the Islanders, even Crossfield, jocularly referred to me as the second 'Wild White Man of Badu'. The first one had lived a century before. An escaped convict from Botany Bay, he had stolen a small boat and sailed it all the way along the eastern coast of Australia to Badu where he was convinced his destiny lay. Stepping ashore, he immediately sank his knife into the heart of the chief and took over the tribe. The warriors thought he was the white reincarnation of Wongai, one of their bravest chiefs. He led massacres on Moa and other islands (there is still a certain enmity between Badu and Moa) until the British governor-general got to hear about him.

A ship was despatched to kill him, and the troopers on board put musket balls through his heart as he charged, screaming, with his warriors down the beach. Then, on the whim of the officer in charge, the troopers sent all the warriors to kingdom come.

It was on Badu that I found out the secret of the green 'pouches' that Lucy and I had seen on Tuin. There was one hanging in the tree outside Ronald's house. They were the breeding bags of large green ants which attacked and bit anyone or anything that invaded their territory. When I had been looking for straight saplings on Tuin's high ground for our house, I had encountered them.

They swarmed all over my body, biting everywhere, and I had to call out to Lucy to help me to scrape them off. Of the several different kinds of ant on Tuin, they were the only ones that bit us; where smaller ants are concerned, 'bite' is the wrong word, for they do not bite but squirt urine onto a person's skin, hence the name 'piss emmet'. The green ants, though, definitely used their relatively large mandibles in their attacks.

My conscience about deserting the people on Badu, when I knew that they obviously needed my services, bothered me greatly.

There was also the self-satisfaction in what I was doing. On two occasions I was taken by dinghy through very rough seas to crayfishing luggers that had broken down and heaved at anchor. There was quite a rewarding thrill when, after toiling in the hot, lurching engine room, I would tell the captain to push the button and away she'd roar.

Crossfield knew of my dilemma – Tuin or Badu – and while Lucy pulled me one way, he sagaciously and craftily pulled me the other. Several times when I had announced that Lucy and I were returning to Tuin on the following day, he would find something very urgent for me to do. I felt at a complete loss about how to confront him and refuse his request. Being a mere male, I allowed Lucy to do it for me.

She told him in no uncertain terms that my book was the reason we were in the Torres Strait and that I had an obligation to my publishers first by living on Tuin, not on Badu.

With a brief 'Get away with you, then', he gave his consent for me to leave. 'Never forget,' he told me, 'I look upon you as my eldest son, and my home will always be yours.'

Eleven

MY FAME HAD spread to Moa and the people there, as well as those on Badu, were soon bringing their engines to Tuin. After consulting with Ronald about rates, I started to charge money for my repair work, giving Lucy and me a choice of which foodstuff to buy. I gave her every cent and she kept it in a housekeeping jar. Several times we crossed over to either Badu or Moa to purchase things from the store.

Sometimes, on Moa, we stayed overnight at the home of the council chief, Oza Bosun, and Lucy didn't much mind that for he was not at all demanding like Crossfield. My work became recognised by the Department of Aboriginal and Islanders' Advancement and I was paid very good money for any council vehicle or engine I repaired. The Aboriginal Development Commission asked me to consider staying on and conducting two-week engine maintenance courses on all the Torres Strait islands with a good salary and air fares paid.

Blissfully happy to be back on Tuin, Lucy was the perfect wife and companion. Sometimes, as Governor of Tuin, I would announce any particular day as a 'Bank Holiday', when Lucy and I would do absolutely nothing, just laze about, discussing philosophy, the meaning of life, past lovers, the stupidity of civilisation, everything and anything. Someone had given me a radio but we turned it on only once or twice for we had no wish to hear of the outside world's troubles and we certainly resented the music. We even sent a letter to Roy Plomley, telling him how stupid we thought his 'Desert Island Discs' programme to be. He never replied.

Often Lucy commented about how she had changed and how she had learned from me the joy and contentment of taking things easy. 'I never dreamed that one day I would be able to sit and do nothing,' she said. Always she ate twice as much fruit as I did, and it was her avid love of fruit that caused her to poison herself for the third time. Ronald had brought us the fleshy fruit of the cashew, with the nut attached. He warned us, before leaving, that on no account were our lips to contact the nut because, uncooked and untreated, it was deadly poisonous with acid as powerful as that in car batteries.

Lucy, determined to get the last ounce of the delicious flesh, allowed the nut to enter the side of her mouth. The chemical action was instant, causing her to cry sharply in alarm and anguish. Burning blisters sprang up inside and outside her mouth, spreading rapidly along her tongue and inside of mouth to her throat.

'Spit, Lucy! Spit!' I cried, seeing what she had done. Commanding her not to swallow saliva, I rushed to the food store and mixed a large quantity of powdered milk with water. Reason told me that alkaline was the best antidote. For an hour I made her gargle to ensure that the blistering didn't travel down to her stomach. I made several mixings and all afternoon she dipped her lips into the bowl, sucking in the soothing liquid and allowing it to float out again. On the following day, she gargled frequently with milk and rubbed zinc ointment all over her mouth. It was several days before the blisters subsided, leaving dead and cracked skin which irritated her for a few days more.

There were two things that Lucy contracted which I didn't: skin fungus all over her breasts and back, and intestinal worms which showed their presence at night with an itching anus that made her claw herself with such intensity that I thought she would cause bleeding. We were both convinced that she had picked up the worms on Badu, for on several occasions we had seen youngsters being allowed to shit on the food table!

One day when I had commented how clothes were more sexy sometimes than nudity, for there was mystery with the latter, she said: 'Let me see what I can find in my suitcase. I'll call you when I'm ready.'

264

In answer to her 'Ready!' many minutes later, I entered the hut and couldn't believe that such a beautiful creature would ever be seen on an uninhabited tropical island! 'It's not much, I'm afraid,' she said.

On her head was the banana-leaf picture hat someone had given to her on Badu. Tied loosely to hang like a ribbon over the brim was a strip of her chiffon. Her cheeks were rouged, her eyes heavily mascaraed and her lips were painted with red lip-gloss. She had used another band of chiffon as a halter-bra to lift her breasts. Sheer, black Dior stockings encased her long legs almost to the tops of her thighs where they met eyebrow-pencil drawings of suspender belt straps which disappeared beneath the legs of the pink nylon knickers.

'Oh, my goodness!' she gasped. 'I don't think anyone has ever taken my knickers down so fast! What about the mystery?'

And that was how Millicent Farquharson, debutante and socialite, was born, together with a sensual and fascinating game. Wearing my linen safari jacket and straw hat, and a dog-collar made of bandage, I would play the role of the vicar, calling on her for tea.

I would knock on the outside of the hut and Millicent would call: 'Come in, vic-cah!' I would enter and sit next to her, shyly, at the table. 'Tea?' she would ask, sweetly and archly, at the same time accidentally rubbing her stockinged foot against my calf. 'Oh, I'm terribly sorry,' she would say, remove her foot, and pour me a cup. Sometimes her hand would inadvertently slip off the table and on to a straining Member for Tuin West, and once more she would apologise, remove her hand and chat on about church affairs and the subject of my next Sunday's sermon. All the time she would be crossing and uncrossing her legs and making sensuous movements with her hands, eyes and lips. Lust often overcame me to a point where I couldn't play anymore.

'You're not acting!' she'd pout. 'That's half the fun of it.' And once again I would assume the role of a rather shocked incumbent.

Sometimes in the afternoons we would just sit in the cool of the hut, and I began to teach her to play cards. Quite frequently, one of the many geckos which lived in the roof would shit on her left breast. 'It's always my left breast,' she would say, laughing, 'and why does the little devil always pick on me?'

'Because it likes you,' I said.

She kept the hut spotlessly clean and ticked me off for leaving anything not in its rightful place and I would pretend to groan and say: 'Fancy the Governor of Tuin being henpecked!'

One afternoon we found that the roof had a new resident when a broad, flat head with piercing eyes and long greeny-coloured neck poked itself through between the roof and the wall.

I was reaching for the rifle when I noticed it had front legs with very long sharp claws. It wasn't a snake but one of the poisonous goannas. But it was only interested in the geckos and took no notice of us, so we took no notice of it, either; thereafter, it always lived in the roof.

Although Lucy said she had seen many more, I saw only two snakes on Tuin. The first one, which crawled into camp, I took no chances with and cut its head off instantly with the machete. We kept it for Ronald to identify and he said it was a baby python. Some weeks later when another one had crawled out from under the bush, I stepped on it just as Lucy saw it and screamed, causing me to leap into the air! Recognising it as another baby python we just let it be.

Sometimes Bronzy and Oscar gave minor heart attacks to our visitors by crawling on to their bare feet as they sat sipping tea! 'Snake!' they would cry, leaping up and away.

With our present relationship I don't think I could be blamed for hoping, more and more, that Lucy would stay with me after the year was up, and the days, weeks and months were flying by to that inevitable time when she would have to say yes or no. I think I knew, though, what her answer would be. I simply didn't want to acknowledge it. Many times she had said she would return to England when the year was up.

May arrived, and with it full cycle. No more did we see the variegated blazes of flocks of parrots sweeping through the trees. The black and white pelicans were back and so was the passion fruit.

In all that time, Lucy had never had a period. She began to complain of stomach pains and developed a vaginal discharge. I took her to Badu and Janine immediately radioed for the helicopter. Without Lucy, I felt empty, alone and miserable. I phoned TI hospital that evening from Badu's only telephone, a public call box. A weak-voiced Lucy told me that her IUD has caused a womb infection and she would be hospitalised for about a week. I told Crossfield and Teleai the news, adding: 'I miss her terribly. God only knows what I will do if she goes back to England.'

'Perhaps she no go,' Teleai consoled me. 'You proper lovebirds, and I know Lucy care for you a lot.'

I was given a lift in a dinghy to fetch Lucy from TI a week later. She looked drained and incredibly white. I was so overjoyed to see and to be with her again that I bought a magnum of champagne and we sat in the park, drinking it. 'Here's to us and to Tuin,' I said, raising one of the two plastic cups the bottle shop had given to me. 'To us and Tuin,' she replied.

'I've really missed you, Lucy,' I went on. She put down her cup and looked me straight in the eyes. 'I have made up my mind what I am going to do,' she said, and fear clutched me. 'I am going back to England.'

The pronouncement, although half-expected, was like a death sentence. 'You certainly know how to spoil any occasion,' I said, bitterly.

'I am sorry, but I am going.'

'Why, Lucy? Why?' I demanded. 'We have ten times more relationship, foundation and formula for happiness together than most married couples ever know, let alone start off with. In fact, some would give their right arms to have our comradeship and love.'

'I can't have children by you,' she said.

Instantly I was concerned for her health. 'There's nothing wrong with me,' she said, hurriedly. 'I'm all right down there. It's just that you have had too many children, too many relationships. I don't want to be the end of a line. As it is, I feel secondhand because of our damned marriage. Don't you see, you've *had* your life. A good one, too. You've done exciting things, been to exciting places – now I want to do and see those things, probably more.'

'Why can't we see them together? I may be fifty-two but I'm extraordinarily fit and healthy and everyone puts my age at ten years younger.'

'Can't you understand?' she replied. 'No matter how young you look or how fit you are, you are still fifty-two. People have commented about your being twice my age.'

'My God!' I said. 'What an ego! And you are so impressionable!'

'Look at it this way,' she said. 'Say you reach seventy – eighteen years from now – I'll be forty-four when you pop off and I'll probably have two or three children. That's a very bad age for a woman to be left on her own, Gerald.'

'I see that,' I said, dejectedly. 'Don't worry, I will never ask you to stay again.'

And I never did in the three further weeks we were together. On our last night, I repeated what I had suggested several times: 'You are a brilliant writer, Lucy. I'd like to give you the book. Why don't you write it as Mrs Robinson Crusoe?'

Her answer was always the same: 'It's your book – you write it.'

'Will you ever come back to me?' I asked, on that awful yet sun-bright morning when we were travelling to Badu.

'I don't know. I think not. But I'm so terribly mixed up and unhappy.'

A tattooist's needle could never have left a clearer,

more indelible picture in my mind of my last sight of her – and I know that that picture will be with me until I die.

When the agony-filled time came for saying goodbye, I could not touch her for fear of crumbling in a bereavement of tears. I just looked at her blankly, and used the Australian expression of departure. 'See you later,' I said. Then I stormed off to the house of an Islander who always had wine, to get well and truly pissed and drown my intense grief.

Crossfield drove her to the tiny airstrip. I vowed I wouldn't look when the Landrover passed the window. But I did. She sat with her luggage in the trailer, a picture of wispy-haired loveliness and distraughtness – a cameo of beauty in her blue sarong. Then my First Lady of Tuin had gone forever . . .

I heard later that she had cried in the plane all the way to TI. My flood of pent-up tears didn't flow until a week later when I went over to Tuin to get our medical box. I resolved to be brave. I stood in the camp and she was there – everywhere. So was her call of 'Dah-ling!' God, how I would have loved to have heard it. I looked at the sand-spit and saw a golden-tanned girl performing callisthenics. But still the tears didn't flow.

Mr Poo brought the release as he fluttered into camp and stood upside down in the bush.

'Oh, you bastard! You bastard!' I cried, directing the expletives at both the bird and Lucy. Then my tears were streaming, and a lump in my throat gagged me as I tried to sob mournfully to Tuin that our lovely First Lady was deceased.

Epilogue

WARM, MISTY, sea-blue eyes met mine over the top of her tilted champagne glass. The party's crowd had momentarily divided like Moses's Red Sea, then swirled back to make an island of me and the two people I was with. Excusing myself, I weaved through the tipplers. She had done the same. 'Hello!' she said. 'I'm Jill Levison.'

The ensuing three months I lived with her in her Coogee Beach bungalow throughout Sydney's dry, nostril-burning summer should have rid me of the clinging ghost of the First Lady of Tuin. But they didn't. And I loathed myself for not being able to return Jill's love while I was eating her food and sharing her bed.

Raven-haired and golden tanned, she was thirty-three – yet her breasts stood firm and curving on Tamarara's topless beach. 'I can't understand why no one has taken you to the altar,' I said. 'You must have had plenty of offers.'

'I have never found anyone I've wanted to share my life with,' she said, simply.

She was ex-public school and a member of one of Britain's oldest families, descended directly from Scotland's Black Douglas. 'You have the most exquisite feet and straightest, shapeliest legs I have ever seen,' I complimented her further.

'That's my aristocratic blood,' she replied, completely devoid of affectation.

'I bet you looked terrific in your air stewardess's uniform,' I went on.

'Mummy was terribly upset when I left textile designing for that,' she said, 'and was very pleased when I went back to it.'

Jill fell in love with me yet hated me for not being the

man I would have been had I not been constantly haunted. It was the ghost of the First Lady of Tuin that had forced me to leave the Torres Strait after three work-filled and miserable months, despite the friendliness of the Islanders. 'Please come back,' they had said when I left. 'This is your home.'

On Australia's mainland I drifted, repairing tractors and other engines as I went. In the far south of Victoria there was one afternoon when I had been handed a twelve-bore and cartridges and asked to get a few rabbits for dinner. I'd knocked over five and was leaning over a five-barred gate, the spring sun warming my back. Suddenly, my First Lady was leaning over the gate with me. 'Oh Christ,' I moaned. 'When is this misery going to end?'

I also carried with me another 'ghost' of Tuin – a tropical ulcer on my left leg that refused to heal. Periodically, ulcers had appeared on my legs after the initial lot and were quickly killed by more courses of antibiotics. Lucy had also been forced to take several courses. But this particular ulcer wouldn't go. Doctors in Brisbane and Sydney treated it without apparent effect – the second doctor advising me to eat plenty of fruit and vegetables. Then a farmer told me: 'Try this.' It was a tin of cattle ointment that guaranteed cures for rotten hooves, ringworm and warble fly, among other things. I daubed it on. The ulcer – plus a six-inch circle of skin – died! It had lived for seven months.

I sought an old friend in Sydney who, seeing my torment, took me to the party where I met Jill. But try as she might she could not perform an exorcism. In the end, she told me to leave.

I crossed the broad waste of the Nullarbor Plain and arrived in Perth. I did casual sub-editing for the *Western Mail* and repaired an engine or two in Fremantle Harbour. When I had sufficient money I flew to London. In the two and a half years I had been away, my three sons had become men, each well over six feet in height. 'Hello, Dad,' they said, hugging me. And I knew I was home.

Gradually, I found that memories of Jill were ousting

the First Lady of Tuin from my mind. Jill had always suggested that we spent Christmas 1983 at her family home on Anglesey. It was during the second week of December that I thought to myself: 'I wonder . . .?' Directory enquiries gave me the Anglesey number. I rang it. Her mother answered. I said I was a friend from Australia; had she heard anything from Jill? 'She's here now,' was the reply.

It took me four and a half hours to drive to Cemaes Bay – and I was able to return the love that was waiting and to be completely captivated by those warm, misty, sea-blue eyes.

May 1984
Rowden Paddocks
Bromyard
Herefordshire